The Dinosaur Dealers

The Dinosaur Dealers

john long

ALLEN&UNWIN

First published in 2002

Allen & Unwin
83 Alexander Street
Crows Nest NSW 2065
Australia
Phone: (61 2) 8425 0100
Fax: (61 2) 9906 2218
Email: info@allenandunwin.com
Web: www.allenandunwin.com

National Library of Australia
Cataloguing-in-Publication data:

Long, John A., 1957–.
The dinosaur dealers: mission, to uncover international fossil
smuggling.

ISBN 1 86508 829 3.

1. Vertebrates, Fossil. 2. Smuggling. I. Title.

560

Map by Ian Faulkner
Typeset by Midland Typesetters, Maryborough, Victoria
Printed by McPherson's Printing Group

10 9 8 7 6 5 4 3 2 1

Contents

Prologue   vii
1   Broome Dinosaurs   1
2   Violation   7
3   Reopening the Case   18
4   Investigations in Eastern Australia   29
5   London Calling   46
6   Undercover in Hamburg   59
7   In Frankfurt   73
8   Denver, Utah and South Dakota   85
9   Dragon Bone Sale   106
10   The Fossil Fish Capital of the World   125
11   Fossil-related Crime in South America, India and Africa   137
12   The World's Largest Fossil Fair   154
13   Back to Australia   178
14   The Future of the Fossil Industry   187
Epilogue: A Personal Story   200
Appendix: How to Check if that Fossil is Legal   207
References   213
Acknowledgements   219

Prologue

The theft of rare dinosaur footprints in late 1996 from an isolated beach near Broome, in the far north of Western Australia, sent shockwaves through the peaceful world of palaeontology. The prints were thought to be the only known good trackway of a stegosaur in the world, and the only evidence for this dinosaur family having existed in Australia. The theft so infuriated the local Aboriginal peoples that the Elders threw a curse upon the perpetrators. Never before had such a site, sacred to palaeontologist and Aboriginal alike, been so publicly violated. The crime made front-page news in the *Australian* newspaper on 16 October, and was reported on major news networks around the world. Although local police conducted a thorough investigation into the crime, no firm leads were established and the case was left unsolved. Then, in late 1998, another stolen fossil dinosaur footprint from Broome came into the public eye. This time the thief, a local man by the name of Michael Latham, was caught. He received two concurrent sentences of two years, for the thefts of fossilised human footprints from a remote site in the Dampier Peninsula, and a single large dinosaur footprint from near Broome, in addition to seven years for drug-related charges. In pursuing their investigations, however, the police were unable to link this crime to the first theft of the rare stegosaur prints.

Perth filmmaker Alan Carter read about the thefts of the Broome dinosaur footprints, and had the idea of making a documentary about the case. He approached me to see if I would help him investigate the whereabouts of the missing tracks. Had they left the country, destined for some wealthy private collector's house? Or were they still hidden in a backyard garage in Broome somewhere? Alan's film would also be an excellent vehicle in which to explore the whole issue of fossil site protection and illegal fossil trading. Would I like to be involved in the project? As a palaeontologist with the Western Australian Museum, my brief is to study the fossils of the State. Naturally, I wanted to help recover the stolen prints, so I readily agreed.

Our first step was to contact Sergeant John Yates, of the Western Australia Police Force, who headed the original investigation at Broome in 1996. We then asked Wyoming lawman Sergeant Steve Rogers, a specialist who fights fossil-related crime in the USA, if he would help us. Combining their specialist knowledge with my background in palaeontology, we set off to dig for more clues in Broome. Little did we know that our investigation would lead us on an international hunt for the specimens—to Germany, London, the United States and China. Along the way we would explore the issues of fossil site protection, fossil legislation, fossil export regulations, fossil smuggling, fossil poaching from government lands and the fossil fraud industry.

I want to use this information to try to inform governments of ways in which they can formulate better protection for their fossil sites and, in particular, to assist in the formulation of local legislation in Western Australia. But my main aim at the time was the same as Alan's—to try to recover the stolen dinosaur footprints and return them to their local custodians.

Fossils are a multimillion-dollar business worldwide. One specimen alone, a *Tyrannosaurus rex* skeleton, recently sold for over US$8 million. Where big money is involved, often so is big crime. We never expected our investigation to open up such a can of worms, but it did, as you will see.

Dates and places mentioned in this book are mostly correct. In some cases, however, they have been changed for legal reasons.

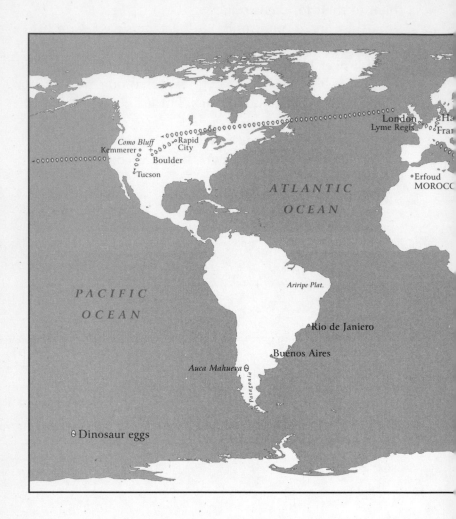

London
Lyme Regis
Ha
Fra

Como Bluff
Rapid
City
Kemmerer
Boulder
Tucson

Erfoud
MOROCC

ATLANTIC
OCEAN

PACIFIC
OCEAN

Ariripe Plat.

Rio de Janiero

Buenos Aires

Auca Mahueva

Patagonia

Dinosaur eggs

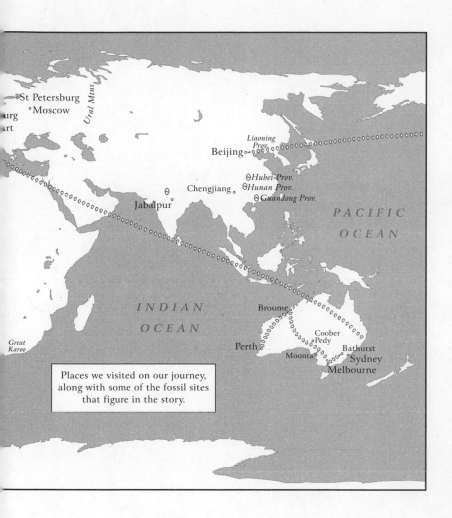

St Petersburg
°Moscow

Ural Mtns

urg
rt

Liaoning
Prov.

Beijing

⊖*Hubei Prov.*
⊖*Hunan Prov.*
⊖ *Guandong Prov.*

Chengjiang

Jabalpur

PACIFIC
OCEAN

INDIAN
OCEAN

Broome

Coober
Pedy

Perth

Moonta

Bathurst
Sydney
Melbourne

*Great
Karoo*

Places we visited on our journey,
along with some of the fossil sites
that figure in the story.

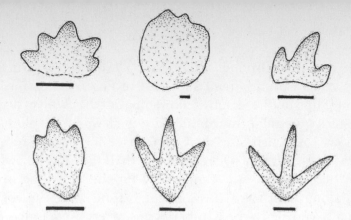

Broome Dinosaurs

1

The sleepy north-western town of Broome is situated on the aquamarine waters of Roebuck Bay and, unlike anywhere else in Western Australia, has water facing both east and west of the township. Broome is a thriving tourist town, popular with foreign backpackers and aristocratic visitors alike. Many come to Broome by discount airfares or on holiday package deals, others arrive by private jet or luxury ocean-cruising yachts. Broome is the centre of Western Australia's pearling industry, and the gateway to the whole Kimberley region for any tourists wanting to head north along Highway 1.

Sketch showing some of the different types of Broome dinosaur footprints. *Top row, left to right:* ?stegosaur handprint (21 cm wide); sauropod footprint (80 cm long); *Wintonopus* foot (17 cm wide). *Bottom row, left to right:* ?stegosaur footprint (25 cm wide); theropod footprint (46 cm long); *Megalosauropus* footprint (52 cm long).

I first visited Broome in August 1986 after a gruelling five weeks working fossil sites east of Fitzroy Crossing. On taking up my research fellowship at the University of Western Australia, my immediate goal was to explore the famous fossil fish sites at Gogo, east of Fitzroy Crossing. I had a prestigious Queen Elizabeth II Fellowship, which provided me with a decent salary for the first time in my life, along with some basic research funds, yet the cost of actually getting to the Kimberley from Perth in a four-wheel drive was still prohibitive. Finally, armed with a large research grant from the US National Geographic Society, I purchased my four-wheel drive and organised a field expedition. It was an ordeal, as our team of three people spent some five weeks working the Kimberley sites, which span an area of some 100 square kilometres. We amassed a large collection of superb fish fossils, several of which later turned out to be species new to science. I was ecstatic with our success. So, as a reward for all my hard work, I decided to bring my wife and young children up to Broome to meet me for a week's holiday.

At the time my wife's uncle was working at the Broome airport and he was able to get us some cheap airfares. He also set us up in a caravan on the site of the old stables near Gantheaume Point. From there we were able to walk to the beach by day and savour the balmy nights in town. We fully enjoyed the hospitality of the local people. I remember that week well, as it was topped off by us watching the grand parade of the *Shinju Matsuri* festival, Broome's biggest event and most popular tourist attraction.

Broome was trying hard to develop into an international tourist destination under an umbrella of plans held by Lord Alistair McAlpine. Ex-Tory party treasurer, McAlpine first visited Broome in the summer of 1979 and, falling in love with the place, returned to buy a house there in 1981. From there he started developing Broome, seeing its potential for international tourism. He built the luxurious Cable Beach

Resort, opened a zoo and, at the time I visited, was trying to set up an international airport. His ideas were indeed on a grand scale, and opened up a lot of commercial opportunities, but somehow neither the locals nor the State government saw the same potential as McAlpine, and the international airport did not succeed. Nonetheless, McAlpine was well known around town as the colourful, slightly eccentric British Lord who had transformed Broome from its sleepy beginnings. He was an avid collector of art, artefacts and anything odd and unusual, and a great supporter of the arts.

During that week in 1986 we tried to visit another of Broome's tourist attractions, 120-million-year-old dinosaur footprints at Gantheaume Point. Unfortunately they could only be seen at very low tide, so we never actually saw them. I must point out here that despite my palaeontological training, in those days I had no real interest in undertaking any research work on the Broome dinosaur footprints as my focus was on the early evolution of fishes.

Broome's dinosaur footprints were reportedly found in the mid-1930s by girl scouts walking along Gantheaume Point. In the late 1940s, the then curator of the Western Australian Museum, Ludwig Glauert, wrote a brief report on their discovery which was eventually published in 1952, first alerting the scientific community to their existence. In the mid-1960s, a well-known American palaeontologist, Ned Colbert, teamed up with Western Australian Museum palaeontologist Duncan Merrilees to make a detailed study of the footprint site at Gantheaume Point. They measured, photographed and cast the trackways and erected a new type of dinosaur footprint name for them, *Megalosauropus broomensis*, meaning 'the *Megalosaurus*-like foot from Broome'. The prints they studied comprised a couple of trackways moving in different directions. The large three-toed prints measured up to 37 cm in length, suggesting that the track maker was a predatory dinosaur of moderate size, maybe only five or six metres long. Colbert and Merrilees published their findings in the *Proceedings of the Royal Society of Western Australia* in 1967. This was the first detailed study of dinosaur footprints from Australia, and the first record of dinosaur fossils from the western half of the Australian continent (Colbert & Merrilees 1967). They did not identify or locate any other fossil footprints in the region, so for the next twenty years no further research was done at the site.

In the late 1980s a local amateur naturalist, Paul Foulkes, became interested in the dinosaur footprints. By searching many of the beaches around Broome during extremely low tides he discovered more and more dinosaur footprints. Moreover, he found footprints the likes of which scientists had never laid eyes upon. In late 1989, only weeks after I was appointed as the Western Australian Museum's first Curator of Vertebrate Palaeontology, I had a letter from Paul, telling me about his new finds. I somehow managed to scrape up some leftover field money from within the administration, allowing me to make my first official museum field trip in mid-1990. My mission was to work with Paul and his friends to document the new trackways.

That week was both intense and exciting. Paul showed me all of the new sites around Broome, as far-ranging as Prices Point, some 70 km north of Broome. He had worked closely with the Rubibi group of local Aboriginal people, and had their permission for us to study the prints, which were sacred to them. During that week I was flat out, working with camera, notebook and ruler in an attempt to photograph, draw and measure all of the many different kinds of dinosaur tracks, ranging from those of large three-toed meat-eaters (theropods, like *Megalosauropus* from Gantheaume Point), huge sauropod trackways (from *Brontosaurus*-like beasts), large and small ornithopods (upright-walking plant-eaters) and the most amazing finds of all, the so-called stegosaur tracks. At this time I had no real idea that these last were stegosaur tracks, so I contacted a dinosaur trackway expert, Dr Tony Thulborn, from the University of Queensland. Tony had just published his authoritative textbook, *Dinosaur Tracks*, and was the acknowledged world expert in the field. He suggested that the combination of a five-fingered handprint and stubby, three-toed footprints could only belong to members of the Thyreophora, most likely the family Stegosauridae. We now know that some other armoured and horned dinosaurs have similar finger–toe patterns, but only stegosaurs show the same asymmetric type of hand-(manus) print as the Broome tracks, so we are still quite convinced that our beast was indeed a stegosaur of some kind—the first one ever recorded from Australia, in fact the only fossil evidence that the group was ever in Australia.

I remember visiting the exact site at the end of the week. Paul had entered into delicate negotiations with local Aboriginal leader Paddy Roe for permission for me, as a member of the Western Australian Museum, to visit the site. As the site is sacred to the local people, we were under strict instructions not to remove anything or draw attention to its exact location. I was more than content to abide by these rules, so I made latex rubber casts of the prints, took numerous photos and measurements, and was thus able to document the existence of this extraordinary set of fossil tracks for the first time. Indeed, as my book *Dinosaurs of Australia* was going to press in October that year, I was able to include a black and white photograph of the stegosaur hand and get the information into print for the scientific world (Long 1990). I was careful not to reveal the locality of the site in my book, referring to it only as being 'near Broome'.

In mid-1991 the Western Australian Museum teamed up with the Australian Army to run a joint field trip, hoping to test the limits of army logistics by working in remote field locations. It was my opportunity to target various remote fossil sites in the Great Sandy Desert and Kimberley areas, with a team of 40 personnel, seven new Landrover vehicles and three Unimog trucks. Under the command of Major John Wild, we set off on 3 July. I had invited various other Australian palaeontologists to join us, so Tony Thulborn and Tim Hamley of the University of Queensland arranged to meet us in Broome. The expedition was also the subject of a documentary film entitled *The Great Aussie Dinosaur Hunt*, which went to air on Australian television nationally in late 1992.

When we visited the rare stegosaur track site near Broome, Tony Thulborn and I successfully cast some of the stegosaur tracks during low tide. We also found a loose block of rock on the beach that showed some of the best stegosaur hand- and footprints in association. During my previous investigation of the site with Paul Foulkes in 1990, Paul had informed me that we had Paddy Roe's permission to borrow the loose slab on the beach and make further detailed studies of it back at the Museum. At the time I had declined the offer. This time, with adequate manpower, and after discussions with Western Australian Museum anthropologist Peter Bindon, we decided to take the block back to Perth so that a complete cast of its surface could be made. In late 1994, after the block had been cast, Tony suggested that we should now return the block to the beach in Broome so as to maintain faith with the local Aboriginal people. I immediately had the block freighted to Broome and Paul and his friends returned it to its exact place on the northern beach, so that it remains within the context of the entire dinosaur trackway at that site, and is an important part of the local Aboriginal Lurujarri heritage trail.

The next time I heard about the famous stegosaur site was in October 1996. News of the dinosaur footprint theft came to me via a phone call from the Kimberley Land Council, a few days before the story was broken to the media.

Violation

. . . the study of trace fossils allows us to reconstruct a remarkably detailed picture of dinosaurs as living animals. It provides glimpses of dinosaurs going about their everyday business, sleeping and eating, visiting the local water hole to drink or to forage. There is evidence of dinosaur nesting grounds, or careful nest building, and of parent dinosaurs tending their youngsters. There are traces of plant eating dinosaurs moving in herds through their feeding grounds, evidence of predation, of solitary hunters stalking their prey, and of opportunists and scavengers roaming in packs.

Much of this evidence may be gleaned from a careful reading of the tracks left by dinosaurs (Thulborn 1990, p. 13).

About 110 million years ago, a stegosaur left its footprints in the sands near Broome. In 1996, some of these fossils were stolen.

Dinosaur footprints are more than just fossils formed from the impressions left by the feet or hands of a dinosaur. The preceding passage emphasises just how much we can learn of dinosaur behaviour from the study of their trackways. Dinosaur bones tell us what a dinosaur may have looked like, but dinosaur trackways tell us how these fascinating beasts moved, rested, stalked, how fast they walked or ran, how they looked after their young. The most important thing to stress here is that each dinosaur footprint belongs to the trackway sequence it is found in. To remove one or two takes away precious information from the complete data set.

As a palaeontologist who often gives public talks about dinosaurs, a question I am often asked is, how are dinosaur footprints preserved? Most people think, rightly, that if you walk along a beach and leave nice tracks, the waves, wind or rain will soon wash them away. So just how do dinosaur footprints get preserved as fossils? The answer is, only in cases where environmental conditions are perfect. The beautiful trackways at Broome and Winton, in Queensland, are examples where dinosaurs have walked along a river flat composed of a mixture of fine sands with an appreciable amount of mud and clay. If the climate is seasonal and a long dry period ensues, the trackways then bake in the sun and are temporarily hardened. Wind-blown silts and fine sands may then bury the trackways, protecting them from further erosion, as can a gently flowing sheet of water. Eventually, after burial by many more layers of sediments, the layer containing trackways becomes transformed by pressure and heat, forming chemical cements that bond the sand grains into a much more durable sandstone rock. The fossil footprints are now locked in as forms within a solid lithic layer. After millions of years of crustal movements, these layers eventually find their way to the Earth's surface, and gradual erosion once more exposes the dinosaur footprints.

In many cases, as with the Broome prints, the exact layer which was depressed by the dinosaur's foot may not be shown on the surface, but instead we see the underprint (sediment layers depressed by the dinosaur's weight which lie below the ground surface) or overprint track outlines (sediments that have infilled the footprint depression and now appear as layers above the ground level where the dinosaur walked). Not only are Broome's dinosaur foot-prints well preserved, but the variety of dinosaur tracks identified there by Tony Thulborn and his co-workers makes the Broome sites among the most diverse collection of dinosaur footprints of that age anywhere in the world.

To study dinosaur trackways the palaeontologist will measure the individual footprints, the stride and pace length, and the angles between each step, calculate the hip height of the animal and estimate how fast it was moving. In rare, extremely well-preserved cases you can even see skin impressions of the underneath of the dinosaur foot. Also, inasmuch as dinosaur skeletons can be identified by their bone morphology, dinosaur footprints can be identified down to individual dinosaur families by the proportions of their feet, type of handprint (if associated with the footprint), numbers of digits, shape of the pads on the foot, and so on. The study of dinosaur trace fossils is a whole new field of palaeontology that is giving us wonderful new insights into dinosaurs as living creatures.

To take fossil footprints out of their natural context of a preserved trackway thus destroys vital scientific information. If the trackways are newly discovered and have not been fully studied to evaluate their scientific significance, such an act is, in effect, a violation of world heritage.

October, 1996. The clear blue salt water of the Indian Ocean rapidly ebbs away from the beach, exposing a broad platform of scattered rocks in the dim moonlight.

The man must work quickly, because it won't be long before the water rises again and covers the beach. This area has tides of up to three metres in places, making for dramatic tidal shifts. He checks his surroundings to see that no-one is near as the location is a popular tourist camp site, well known as a good fishing spot. Satisfied that he is alone, he switches on his torch and lays out his tools next to the rock. He marks out a line, then begins to drive metal wedges into the solid sandstone rock. It is harder work than he expected; it takes him a good hour of constant work to drive the metal plugs in a straight line along the rock. Nervously he stops and looks around. The metal plugs form a neat line that fractures downwards, forming a vertical split in the rock. Using broad-headed masonry chisels he then tunnels under the rock layer, finding its plane of weakness in the horizontal bedding plane.

A few hard whacks with the sledgehammer finally free the block. It comes away perfectly, forming an almost straight break. He grunts, heaving the heavy rock up into his arms and, after wrapping it in hessian, carries it about 50 metres back to his vehicle. He quickly returns to gather his tools, then drives away. Sitting in the back of his old pick-up truck is a slab of rock, in the middle of which are two depressions representing the places where, 120 million years ago, a dinosaur left its footprints. Not just any old dinosaur footprints, but exceedingly rare ones. He has no real idea why someone would even want these footprints, but he does know that they are worth a pretty penny, and has lined up a quick sale. He smiles to himself at how easy the job was. No-one will ever miss them, he thinks to himself.

Thursday, 10 October 1996. I receive a phone call from George Irving of the Kimberley Land Council, the body which deals with native title claims for the Kimberley district.

He tells me that dinosaur footprints have been taken from a site north of Broome, and asks if I can help them identify the footprints and provide information about their rarity and scientific value, as well as photographs. After being given more specific details about the exact site and a brief description of the footprints, I confirm that I do have photographs in my files of many of the dinosaur footprints from that site, but to be sure of exactly which ones were missing I need to make some enquiries of my colleagues. I send the Kimberley Land Council a fax expressing concern about loose blocks containing fossil footprints which I knew about at that site.

The theft is reported to the local police at Broome. Sergeant John Yates begins investigations.

15 October 1996. CNN news reports the theft of the dinosaur footprints from Broome. The story is headlined as 'Thieves walk off with sacred dinosaur footprints'.

Below is an extract taken from two news articles which interviewed local Aboriginal leader Joseph Roe and local anthropologist Patrick Sullivan.

'It's a very sacred thing to me,' said Joseph Roe, Aboriginal custodian for the past eight years of the footprints near Broome, on the country's remote northwest coast.

'According to Aboriginal tradition, whoever has taken them has placed themselves in great danger,' he said. 'They might get sick or I might get sick.'

'The offence was punishable by death under Aboriginal law,' he said.

'If he [a thief] comes to face me I will put a spear through him and finish him,' Roe said by telephone from Broome.

Anthropologist Patrick Sullivan, among a party of Aborigines who discovered the theft last Wednesday, said the footprints were part of a 'song line' of sacred sites used in Aboriginal ceremonies.

He said the Aborigines with him were outraged, shocked and horrified to find the footprints missing.

'People responsible for looking after these areas feel that if they [sacred sites] were disturbed that sickness and other kinds of misfortune are going to come upon their communities and themselves, and of a very severe kind,' he said from Broome.

Roe appealed for the thieves to return the footprints, which are registered officially as a sacred site in Western Australia.

The Western Australia state premier Richard Court called the theft callous and 'sick' and pledged tougher penalties and tighter security for fossil sites. He also offered police all government resources to investigate the theft.

16 October 1996. The story makes front-page news in the *Australian* newspaper and is on page two of the *West Australian.* Shockwaves ripple through the palaeontological world, as never before in Australia has a site of such scientific and Aboriginal significance been so violently desecrated.

Over the next few days there was a rapid exchange of correspondence to try to verify exactly which prints had been stolen. I emailed Tony Thulborn, as he had been carrying out detailed research on the Broome dinosaur footprints and was familiar with most of the trackways in the region. I had done some preliminary research on the sites in 1990 and had a fairly good register of colour slides of some of the better, and scientifically rarer, trackways, so I wanted to liaise with Tony about the missing prints. I sent him an email attachment of a photo showing the two hind footprints of possible stegosaurid dinosaur, and he replied at once confirming that, based on the information he had received from Paul Foulkes and the local police, these were the missing ones.

I then got back in touch with the Kimberley Land Council and sent them a colour photograph of the missing prints, together with a copy of my 1990 field notes, outlining in my sketch the full extent of that dinosaur trackway as I then understood it. I also highlighted which prints I understood to be missing. Soon after the police investigation got under way Interpol was called in, on the assumption that such rare fossils could have been whisked overseas to a waiting buyer or specialist trade fair. Sergeant John Yates phoned me to discuss the matter. We concurred that, as only a handful of people knew that the footprints were at that site, the theft must have been a specialist job. Perhaps, we hypothesised, a collector or dealer 'ordered' them specifically, knowing they were rare. Sergeant Yates asked me if I would prepare a witness statement outlining the scientific rarity and importance of the stolen fossils. I agreed, and said I would forward my own photos of the missing prints, together with any other information I had at hand about them.

Yates' initial investigation focused on various locals who had previous convictions for theft of artefacts, or had experience working with stonemasonry, but none of the leads produced any concrete evidence. Local fossil dealers in Perth were also questioned but again, nothing conclusive could be established. Finally, after a few weeks, the case was shelved, unsolved.

News of the stolen dinosaur footprints received world-wide media attention, bringing the whole issue of protection of important palaeontological and archaeological sites to the fore. The Broome site was on an isolated beach, well away from the town. As it was a public area, there was nothing authorities could do to protect such a site from unwanted visitors. Leading politicians vowed to bring in legislation to protect the site, and other significant fossil and archaeological sites (as stated by Premier Richard Court in the *West Australian* on 16 October 1996).

On 7 June 1998 Paul Foulkes, who first discovered the rare stegosaur footprints and alerted me to their existence, passed away after a prolonged battle with cancer. Many think that the worry Paul endured over the vandalism of the site probably contributed to his ill health. His ashes were scattered over Riddell Beach, one of his favourite dinosaur footprint sites.

So the case of the missing stegosaur footprints was unresolved, seemingly unsolvable. Then, in late 1998, further events came to light involving another theft of dinosaur footprints from the Broome region. Was there a connection?

27 November 1998. A man in Broome is investigated for drug-related crimes, and a videotape of a stolen dinosaur footprint along with fossilised human footprints is found after a raid on a cattle station.

The following piece is an extract from the day's CNN report:

> Australian Federal Police have launched an international hunt for a 120-million-year-old fossilised dinosaur footprint, stolen from a sacred site near Broome. Federal agents in Europe and Japan are on the trail of what is thought to be a theropod print and rare human footprints, which are possibly worth hundreds of thousands of dollars. Two West Australian men were charged with stealing the footprints from a sacred site [*x km, here deleted*] north of Broome.
>
> Charges were laid while the country was still reeling from the theft of another priceless Aboriginal artwork from the Tasmanian wilderness, possibly destined for the international black market. AFP agent Russell Northcott said the Australian Customs Service had been called in to work on the case and officials were confident of finding the rare and significant artefacts, believed to have disappeared late last year.
>
> The footprints, including the rare 7000-year-old human tracks, disappeared from secret areas around the artefact rich land around Broome in Western Australia's steamy north-west. They were taken just over a year after Australia's only known set of *Stegosaurus* prints—the one piece of direct evidence that the

dinosaur existed in this country—were hacked from a rock in another sacred Broome site.

I should point out here that all the dinosaur footprints came from the same geological layer, the Broome Sandstone, which was around 120 million years old. The human footprints were from a beach site in the Dampier Peninsula, and dated at around 5000–8000 years old. Several of the media reports of the day misled the public into thinking that the human and dinosaur footprints were found together in rocks of the same age!

The police would later recover the fossils—on 30 December 1998. A local media report on 7 January 1999 quoted Senior Sergeant Geoff Fuller from Broome Police as saying, 'We know there have been attempts to sell it [sic] in Asia, but perhaps because of its size and weight or for whatever reason they've been unsuccessful'.

News headlines around Australia and throughout the scientific world then further misled readers by claiming that the stolen dinosaur footprints (referring to the previous 1996 case) had been recovered. When I was contacted by the police at Broome and could confirm that the recovered fossil footprint was quite different from the ones stolen in 1996, it was clear to me that the two cases were *not* necessarily related. The recovered fossil footprint belongs to *Megalosauropus broomensis*, one of the more common varieties of three-toed carnivorous dinosaur footprints found all around the Broome region.

Michael Charles Latham, 46, admitted having stolen the dinosaur footprint from a site on Roebuck Bay, Broome, along with three fossil human footprints from the northern part of the Dampier Peninsula. A second man was also charged in connection with these thefts: Rodney Illingworth, the manager of Roebuck Plains Station, where the video of the fossil was found after a police search.

Once again I was asked to submit a witness statement outlining the rarity and scientific significance of the Broome dinosaur footprints. Eighteen months later, as the case came to trial, I was asked to make myself available as an expert witness. As the defendant pleaded guilty at the last moment, however, my presence wasn't required. Rodney Illingworth was found not guilty, as we have seen, and Michael Latham guilty.

This was the first case in Australian legal history of a person being sentenced to jail for stealing a fossil. But it wasn't over yet.

21 December 2000. AAP/Reuters report:

A man acquitted of receiving a stolen fossilised human footprint is to be retried.

Rodney Grant Illingworth, 31, had been acquitted in the West Australian Supreme Court earlier this year of receiving the 7,000-year-old footprint from a man convicted of its theft.

But the Crown appealed against the verdict and the WA Court of Criminal Appeal today quashed the acquittal and ordered a retrial.

Illingworth had been tried separately to 47-year-old Michael Latham, who is serving two years' jail for the theft of the human print and a 120 million-year-old dinosaur footprint at Broome in October 1996.

At his trial, Illingworth, manager of the Roebuck Plains Station, pleaded not guilty to receiving the stolen human footprint from Latham.

Broome police had told the court that a video had been found in a police raid on the property showing a person with painted toenails putting their foot inside the fossil.

An ex-business associate of Illingworth, Gregory Travelstead, who once collected artefacts for Broome developer Lord Alistair McAlpine, told the court Illingworth had asked him if there was any interest in fossilised footprints.

Mr Travelstead suggested he contact anthropologist Lindsay Hasluck, an anthropology lecturer at Deakin University. The court was told Mr Hasluck was interested in studying the footprint but could not afford to travel to Broome.

Illingworth was acquitted of receiving the stolen human print after his defence argued the Crown had to prove he had 'evil intent' with the artefact, such as an intent to sell it.

But the WA Director of Public Prosecutions appealed, arguing before the Court of Criminal Appeal in June that someone was in breach of the law if they had possession of something knowing it was stolen and the Crown did not have to prove intent.

This report mistakenly said that Latham had been charged over the footprints stolen in 1996 which, as we know, is not true. He was charged with the theft of dinosaur footprints from the Crab Creek area, Roebuck Bay. Was there a connection between the two thefts? If so, where did the stegosaur prints end up?

The time had come to make further enquiries, but we would need to bring in expert help. An expert in fossil-related crime.

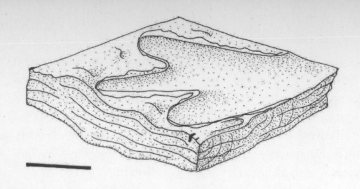

Reopening the Case

3

'Worldwide, the theft of fossils is gaining momentum, with Russia, China, Australia and the United States leading the way,' said Angela Meadows, program manager of cultural property for Interpol in Washington . . . The best way to crack down on fossil stealing is to catch thieves red-handed, but that is difficult in the West. A ranger for the U.S. Bureau of Land Management in Wyoming patrols an average of 3 million acres; Steve Rogers alone patrols about 500 square miles of the Green River Formation, which encompasses thousands of square miles (Associated Press, 8 February 2000).

A sketch of the slab of rock containing a dinosaur footprint which was removed from a beach near Broome in 1998. The bevelled semicircular edges on the right side of the slab are saw cut marks. (Bar scale is 10 cm)

18

When Alan Carter of Alley Kat Productions approached me in mid-2000 to see if I would be able to participate in his documentary film series about the stolen Broome dinosaur footprints, I joked with him, saying that if the police had not been able to solve the case, despite all their efforts, what made him think we could get any further with it? He told me that he had an ace up his sleeve. He had located a top US lawman, Sergeant Steve Rogers from Kemmerer in Wyoming, who was an internationally recognised specialist in fossil-related crime. If anyone had the experience and ability to crack this case, Steve Rogers was that man.

That day I did some Internet research about fossil-related crime. There were several articles about the fossil cop from Wyoming. He had been active in catching poachers taking fossils from government lands in Wyoming and tracking stolen or smuggled fossils from sites around the world.

I phoned Sergeant Rogers and asked him if he would like to be involved in an investigation of the theft of rare dinosaur footprints from the town of Broome, in Western Australia. His voice sounded confident and calm, his speech punctuated by long, deliberate pauses as he soaked in every bit of information I gave him. He asked me if we would be working closely with the police and I replied that we would indeed be working with the local authorities, and I would arrange for him to meet with the policeman who headed the original investigation, Sergeant John Yates. He asked for some more details about the case, and said that he would consider it, if he could get the time away from work and his expenses would be covered.

'But why me?' he finally asked.

'Because you're the best in the fossil crime world,' I replied.

Sergeant Steve Rogers is a fully qualified pilot who can fly anything with wings, also a qualified helicopter pilot. He regularly flies aerial patrols over State-owned lands in

Wyoming. One of his main jobs is to spot fossil poachers who are illegally digging for fossils on such lands. Over the years Steve has been involved in many facets of the illicit fossil-dealing world, from catching smugglers attempting to export illegally obtained fossils red-handed and tracing fossil traders' bank accounts on the suspicion that large sums of money had been laundered through undisclosed fossil sales, through to busting smugglers who have been dealing fossils to fund drug rings. His most successful initiative was Operation Rockfish, which resulted in hundreds of convictions relating to fossil poaching from Wyoming State land, and a number of related arrests because of connections to drugs and illegal arms. Steve's many years in the US police force have exposed him to the whole gamut of serious crime, including breaking up narcotics rings in Mexico, so he was very experienced in the ways of smugglers.

After reading about his credentials and experiences, and speaking to him on the phone, I was convinced that Steve Rogers was the one man who could really be useful to our cause, so I was greatly relieved when he called back to say that he would help us. We arranged for him to fly to Perth as soon as possible.

22 August 2001. I meet Steve Rogers at the Perth International airport, then drive him to his hotel in Fremantle, showing him some of the sights of Perth on the way. This is his first visit to Australia, and he seems happy to be here, away from his normal high-pressure routine. We begin to discuss the Broome dinosaur case that evening.

Steve suggested that we could get some information about stolen fossils from the Internet. His idea was to set up an Internet 'sting' in which we send out messages to fossil collectors worldwide, enquiring about rare Australian fossils for sale. So we sent out some emails enquiring about the availability of certain items, including dinosaur footprints.

It was a real long shot, but it wouldn't do any harm if no-one replied, so we thought we'd give it a go. More importantly, anyone who did reply would more than likely know where and how to get dinosaur footprints.

I gave Steve my complete file on the two stolen dinosaur footprint cases, so he could familiarise himself with it before we headed north to Broome. I had also given him the complete court transcripts of the recent trials of Latham and Illingworth. He said he would look over it all when he had recovered from his jet lag.

I had also been in contact with Sergeant John Yates, who was now posted at Geraldton. He was interested to hear that Steve Rogers would be helping with our enquiries, and he agreed to join us in Broome to start a new investigation. Officially it was still his case, so it would look good for the local police to solve the mystery of the missing footprints.

Saturday 25 August. We depart for Broome. That night we walk around the town to take in the lie of the land. At the back bar of the Roebuck Hotel, amid a noisy bunch of locals, we sit with a couple of beers and discuss the case.

Our course of action for the next few days would be first to make enquiries at the local police station as to what information they had on the case, or any new information at hand, then to visit the scene of the crime with Joseph Roe, Paddy Roe's son. We would also make discreet enquiries around town, in shops and so on, to try to pick up any local rumours about the missing dinosaur footprints. In Broome local rumours are easily started, and have gained the nickname of 'brumours'. Any information, even brumours, could be valuable.

Sunday 26 August. Steve and I visit Sergeant John Yates at the Broome police station and discuss the case with him. He produces the complete police file on the 1996 stegosaur print case, including detailed photographs of the crime site which he took the day it was reported.

Steve and I examined the photographs. The photos showed lines of plug and feathering in the rock. These are vertical lines where a metal wedge has been hammered down into the rock so as to make it split along a certain planar direction. We noted that the uneven sides of the break, where the rock was removed from its bedrock, show that the 1996 theft was not carried out using machinery such as a masonry saw or an angle grinder (as was erroneously reported in some media reports at the time). It appeared to have been chiselled out by hand.

John then showed us a replica of the stolen footprint, recently recovered from the 1998 case. Careful examination of this footprint could give us valuable information about how it was removed, and whether or not the 1996 theft was carried out in the same fashion. This could be evaluated better after revisiting the scene of the 1996 theft. Although it was only a replica, the cast was so well made that Steve picked up some clues directly from the fibreglass sides. He pointed out that the saw cuts forming the straight sides of the block were made by two different thicknesses of masonry saw blades, each clearly discernible. One saw cut went into the block and then stopped, as its projected line would have cut off part of the footprint. This small cut gave us one blade width to work on. Someone had done a quick, clean job, someone who knew what they were doing and had the right equipment, even spare masonry blades, at hand.

That afternoon Alan and John visited Joseph Roe. He had just returned from a very busy weekend of presiding over Aboriginal law business, and was worn out from dancing through all the ceremonies. He briefly talked with them about the case and agreed to take them to the site, but he didn't want me to be present, mainly because he didn't want the Western Australian Museum involved with the site. The Museum's previous involvement at the site had caused some dismay among local Aboriginals, because of the documentary made in 1991, when we visited the site with Tony Thulborn, Peter Bindon and the Perth Logistics Battalion of the Army. Some of the locals believed that the film had focused too much media attention on the significance of the dinosaur tracks, and may have prompted the second theft in 1996. So, not wanting to upset anyone, I agreed to remain behind while Steve, John and Alan visited the fossil site the following day.

Monday 27 August. I stroll around Broome and buy some reference books at the Kimberley Bookshop. I chat to the sales lady there about a booklet on the Broome dinosaur footprints, written by a local author. She gives me the author's phone number so I can follow up with further enquiries later.

That morning Joseph Roe took Steve and John to the site from where the stegosaur prints had been removed in October 1996. Steve carefully examined the rock from which the missing block had been chiselled, noting the plug and feather marks which meant that the slab had been removed without the use of power tools. John had previously photographed the site and agreed with Steve's observations. Joseph explained the site's significance to Steve, saying that under Aboriginal law, the perpetrators of this crime would face severe penalties if they were ever caught.

That afternoon we visited Gantheaume Point, or Minyirr as the locals call it, the site where dinosaur footprints were first discovered at Broome. I explained to Steve and John that these footprints have never been stolen or vandalised because they are so rarely exposed, only being visible at extremely low tides that occur a few times a year. Not many people know where to find the footprints, or how to recognise them if they do stumble across them, so the site is well protected by its very nature. Despite this, nature is also the site's worst enemy. Coastal erosion and the corrosive action of rock-burrowing invertebrate animals are slowly destroying all the dinosaur footprints in the Broome area. The only saving grace of the situation is that as erosion destroys some footprints, new ones are slowly being uncovered.

Tuesday 28 August. We visit Broome airport to look at security. The first thing we notice is how many private planes come and go. It is a very busy little airport, where anyone with access to a private plane or helicopter could easily have transported the stolen footprints out of town to a remote landing spot for courier by ship to another country. Private planes owned by pearl-farming corporations also take off and land on a daily basis.

That afternoon we visited the Broome port, a bustling harbour with an impressively long pier for deepwater anchorage. The port was opened in August 1899, but it wasn't until 1966 that the new deepwater port was built so that boats were not dependent upon tides for access. Half-a-dozen pearling boats came roaring in as we gazed over the emerald-green waters of Roebuck Bay. We were surprised to learn from Port Security that the Broome harbour is the only major shipping port in Australia without constant video surveillance. Our day's efforts have only confirmed our suspicions that it would be very easy to smuggle illegal contraband out of Broome, from both the airport and the port.

Wednesday 29 August. Steve suggests that we talk with Michael Latham, currently in Broome Prison. We try to make contact with a local middleman who knows Latham, to tee up a meeting. In the meantime, we call up Digby Macintosh, who was in Broome Prison with Latham. Steve thinks that it might be worth speaking to him to see what he knows about the case.

Digby invited us to visit him at his home that afternoon. We asked him about what he knew or suspected with regard to the 1996 stolen footprints, trying to fish out what the brumours were. Digby had no hard information, but he suggested that the footprints may have been stolen to order, possibly for a wealthy collector. Another brumour we heard that day, also second-hand, was that the stolen dinosaur footprints are now at the bottom of an elaborate fountain, somewhere in Europe.

'All very nice stories,' Steve said to me as we left Digby's house, 'but nothing hard to go on.'

Wednesday 29 August. Early that evening we go to Cable Beach to think about everything we have uncovered so far, to see if we can draw lines between all the different bits of information. Steve and I are being filmed as the sun sets.

I went for a swim while Steve contemplated the serenity of the dying sun's rays. Later we went to the top of the beach to talk on camera about the investigation. Tourists surrounded us, some watching us and the film crew, no doubt hoping to see someone famous.

It was during this interview that we noticed two suspicious-looking young men, both very well dressed (perhaps too well for your average tourist at the beach) and both paying very close attention to us, in fact, hanging on every word we said to the camera. They were in their late twenties or early thirties, both of slender build, one with long hair tied back in a neat ponytail. While we were filming they turned towards Steve and shot video film of us, putting their handheld camera right in front of Steve's face for a fleeting moment. Our soundman noticed this and discreetly fired off a couple of digital shots.

We decided that we would try out the cafe at Cable Beach, The Sand Bar and Grill, for dinner. Steve wanted to stay behind and rest, so he didn't accompany us. As soon as we arrived, however, we noticed the two men from the beach. They pulled up just in front of our car, then stood leaning against their car, casually keeping an eye on us. As we went in to dinner they disappeared. One of our crew wandered past their car and copied down the numberplate, and mentally made a brief description of the vehicle: a white souped-up Commodore with sticker identifying it as a 'Top Class' rental car.

After a quick dinner, about 9 pm, it was quite dark and there were only a few people loitering on the lawns at the top of Cable Beach. As we walked towards the car park, the same two men suddenly appeared in front of us. They walked slowly to their car, got in and drove away, just before us. We now suspected that they were tailing us, so we decided to wait in our car until they were well out of sight. They turned left into the Cable Beach Club. As we drove past, heading towards Broome, we anxiously watched our rear-view mirror, but there were no headlights behind us. We were almost back at our motel when we spotted them, so rather than let them know where we were staying, we drove off in another direction, leaving a decoy trail to see if we were really being followed. We headed towards the Mangrove Hotel. As soon as we parked the car in the front of the Mangrove we spotted them again, slowly cruising past the hotel to check us out.

Inside, safe at the almost deserted bar, we rang John and reported the incident. We gave him the car details and he promised he would run them through the police computer for us and phone back right away. We waited nervously until John rang us back, about 30 minutes later. He told us that the car wasn't a rental car, nor was it local. Its plate identified it as a Perth-registered car, but there were no details either of it, or its owner, on file. John also checked with local and Federal police, just in case our tails were undercover police agents working on a restricted project, but it was soon confirmed that they weren't. We made our way back to our motel, this time without being followed.

The next day we learn that John has handed the details of the incident over to the Federal police, who have taken over the investigation. Something is not right. Why would a film crew investigating the theft of dinosaur footprints attract so much attention? Perhaps there was a more sinister reason for us being followed. Recent events might provide the explanation.

A few weeks before our visit a large shipment of cocaine, worth about AU$160 million, was washed up in the north of Shark Bay (about 1000 km south of Broome). The Geraldton police investigated the crime scene and suspects were charged. John told us that the shipment was originally destined for Broome, so we reasoned that there must have been some pretty major drug criminals hanging around Broome in the last few weeks, waiting for their merchandise. The presence of an American lawman in Broome might have sparked their interest in us.

Lord McAlpine was mentioned in the transcripts of the Latham and Illingworth trials. A Mr Travelstead was the man supposedly approached by Illingworth to see if he had any interest in the stolen (that is, the 1998) fossil human and dinosaur footprints; Travelstead was known to have had business dealings with Lord McAlpine. In reading Lord McAlpine's biographical memoir, *Once a Jolly Bagman*, I noted that he did indeed own an antiquities and collect-ables shop in London. In discussing visits to his shop by the well-known Australian artist Sidney Nolan, he describes the interior as:

> a cabinet of curiosities, containing everything from six-legged lambs to dinosaurs' eggs, from Renaissance bronzes to Roman silver, jewellery, weapons, mystic objects, ethnographical material,

minerals, stuffed birds, skeletons, paintings and carvings (McAlpine 1998, p. 106).

Lord McAlpine had, therefore, had some dinosaur eggs in his collection of exotica. It would be natural for such a collector and trader in unusual items to have an interest in dinosaur footprints, but again, this is just supposition. If we could find Lord McAlpine and interview him, we could surely clear up any false trails.

Friday 31 August. At 8.30 am we discreetly slip out of Broome and head to Perth. At 2.40 pm we fly to Melbourne, where we will spend the night at an airport motel. Our plan is to meet with Tom Kapitany, Australia's largest dealer of fossils, at his Keysborough shop the next day, to get his professional advice on the types of people who would want a dinosaur footprint, and what sort of price such a rare and exclusive item might fetch on the international fossil market.

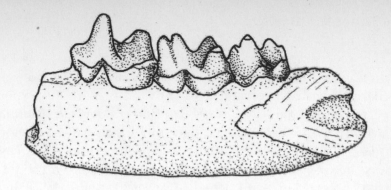

Investigations
in Eastern Australia

4

November 1991. The Perth bust [at the International Airport, of fossils ready to be transported overseas] was followed in November by Federal Police searches of 13 residential and business addresses in Western Australia and South Australia. About 700 fossils are in police custody. Brian Swift, Federal Police national liaison officer, says 'If the suspected value of the fossils is confirmed, it could end up being the largest dollar value fraud case in Australian history' (Bunk 1992, p. 56).

Sketch of the opalised jaw of *Steropodon*, the first fossil mammal from the age of dinosaurs found in Australia. The specimen is just under 3 cm long.

The preceding quote refers to a well-publicised case of three fossil traders operating in Australia in 1991 who were charged with exporting fossils overseas without export permits. At that time the *Protection of Moveable Cultural Heritage Act 1986* stipulated that only fossils valued at more than AU$1000 needed to have export permits. The case was the first such incident in Australian law, and it came to trial in Perth in April 1997, after the Federal Police had spent years tracking down the specimens and gathering evidence for the trial, in an investigation christened 'Operation Bud'. The specimens the fossil traders were going to export (and presumably sell) overseas included well-preserved Gogo fish fossils and exquisite crinoids from Jimba Jimba Station (both sites in Western Australia), as well as rare Ediacaran specimens (the oldest body fossils of primitive organisms on Earth) from the Flinders Ranges and beautiful crinoids and shells preserved in opal from Coober Pedy in South Australia.

The specimens that had already been exported without legal permits and sold overseas were eventually traced by the Australian Federal Police to museums in Japan and dealers in Germany. Acting under a diplomatic request from the Australian government, German police were able to confiscate all of the Australian material (more about this later, when we visit Germany and speak to the policeman who carried out the raids). The Japanese museum which had purchased the large slab with an Ediacaran sea-pen fossil (one of the world's best Precambrian fossils, about 600 million years old) didn't want to be involved in a diplomatic incident so readily agreed to give the specimen back to Australia. In mid-2000 it was repatriated to the South Australian Museum, where it is now on public display.

I was called by the Federal Police, along with other palaeontologists from around Australia, as an expert witness. I remember the court proceedings well. At one stage the defence lawyers were questioning my authority about the Gogo fishes, trying to argue that the defendants had collected them as 'limestone samples' and, as limestone was a commercially mined mineral resource, they were quite within their rights to do so. I recall answering back that Gogo nodules were high in mud and silt content and therefore had no commercial limestone value. The defendants had only collected nodules with fossils in them, specifically ones with mostly fish fossils.

This landmark trial resulted in fines of AU$50 000 being given to one of the dealers (later revised to AU$35 000) and forfeiture of some of the fossils. The legal ownership of the illegally collected fossils was in dispute for some time, but eventually (some years later) a ruling was handed down that the defendants had acquired some of the specimens without breaking any laws, and should be given those specimens back. Today one of these men, who prefers not to be named, works as a curator of a large private prospector's mineral collection, as well as continuing to trade in rare minerals. I see him occasionally and he has prepared fossils for us at the Western Australian Museum, at no charge. In retrospect he sees that I bore no malice towards him in testifying at his trial; the only issue was that he had not filed export permits. In fact most of the specimens in the case would have been granted an export permit as they were specimens well represented in museum collections. A few of them, though—namely the Ediacaran material and Gogo fish specimens—were quite unique and would not have been given permits.

Such specimens need to be sold to an Australian museum, or donated under the Tax Incentives for the Arts scheme (which I will explain later). Unfortunately, at the time Australia did not have adequate funds for the purchase of heritage items (we do have a small fund today for such purposes, but it is not easily accessible). This results in a gridlock situation, where the dealer has a fantastic fossil that the local museum wants, but the museum lacks the funds to buy the specimen. The dealer needs to make his living, so rather than wait for philanthropists to step in (a rare occurrence), or for funds to appear miraculously, he risks illegally exporting the fossils so that he can sell them on the international market.

This trial marked Australia's passage into the world of fossil-related crime. It also brought up many issues about how we value fossils, whether by their monetary or their scientific value, and whether or not some fossil sites should be completely off limits to commercial collectors. Many of these decisions come down to individual States. These issues are difficult to resolve and the fossil trading industry in Australia is flourishing, despite the concern of many scientists around the country.

To understand the extent of the international trade in fossils, one needs only to talk with some of the big commercial dealers and see their elaborate preparation facilities. Melbourne dealer Tom Kapitany is one of Australia's largest commercial fossil dealers, so we decided to pay him a visit and ask him about the possible whereabouts of stolen fossils such as the Broome dinosaur prints.

It was a rainy Saturday morning in September when we visited Tom Kapitany's shop, 'Collectors Corner Gardenworld', in Springvale. Originally a garden centre, the business expanded to include the Kapitany brothers' private collections of fossils, which were put on public display. Gradually they began acquiring local fossils and importing exotic specimens from around the world to sell as collectors' pieces. Today Tom Kapitany's business includes a new, million-dollar warehouse and state-of-the-art workshop for fossil preparation. He is Australia's largest dealer in fossils, and supplies many of the smaller rock and curiosity shops around Australia. His shop sells some specimens valued in excess of AU$20 000 from Australian sites, for example, wonderful opalised belemnites and large slabs of Permian crinoids from Western Australia. The bulk of Australian fossil sales are at the bottom of the market, the sort of object that the average private collector can afford (in the range of AU$10–$100), although the real enthusiast can sometimes find rarer objects, valued up to AU$1000.

Tom has taken out several exploration leases on fossil sites in Australia in order to mine them for specimens to trade on the local and overseas markets. This procedure is legal, although the terms of 'exploration leases' are for *sample* collecting only. There is nothing within mining law that gives any protection to rare or new species of fossils that could be discovered through such practices. The only way that scientists with a research interest in these sites can become involved is with the dealer's or landowner's permission.

Tom and I chatted about his own philosophy of the Australian fossil trade, and his opinions about our laws. Although largely happy with the way things operate here, he sometimes has difficulties getting export permits within short time frames, and readily blames the scientists who act as examiners for the government for holding up the process. As one of these expert examiners, I responded that some-times we are away on work (field trips or overseas), or leave and, as Australia has too few expert scientists capable of doing the job, an examination has to wait for our return. Furthermore, we are not paid or recompensed by the government for doing this highly specialist work, so why should it be rushed through when we have so many other pressing matters at hand?

We agreed that there needs to be a better system for the legal export of fossils for traders, one that also ensures that experts are properly paid for their time. Part of the job of an expert examiner is to make sure that the specimens in question are what they are claimed to be on the export form. In most cases this is simply not possible as, naturally, the Federal government will not pay for an expert to be flown interstate to look at someone's fossils just to confirm what they are. Yet the system has worked so far, albeit somewhat clumsily, largely operating on a high degree of trust between the dealers and the expert examiners.

Tom showed me around his shop. The extent of his stock was truly astounding: aisle after aisle of metal shelving, packed with hundreds of boxes of trilobites from Morocco, Wyoming fossil fishes, fossil leaves and petrified polished wood, large slabs of sandstone with crinoids from Western Australia, shelves full of dinosaur eggs from China and Patagonia and beautiful opalised fossils from Coober Pedy and Lightning Ridge. In addition, he sells and trades a huge variety of mineral specimens, sea shells and eclectic works of art such as stone carvings, the sorts of things that collectors of the odd and unusual would appreciate.

As we have seen, the main buyers of important or scientifically rare fossils in Australia are often the State museums, provided that all the legal protocols are in place. My own institute, the Western Australian Museum, has on occasion purchased such fossils, mainly specimens originating from Australia. One specimen we bought from Tom was a partial skull of a Triassic amphibian, about 220 million years old, from the north of Western Australia. Although not a particularly attractive specimen as most of the bone was weathered away, leaving only impressions of the back of the skull, Tom recognised it for its scientific value and offered it to us for AU$200 (his cost price). I rang my colleague Dr Anne Warren at La Trobe University, who examined the specimen and confirmed that it was a good skull and that it had come from the Blina Shale unit of the Erskine Ranges of the Kimberley district. It had been offered to Tom through a private collector. We arranged for Anne to pick up the fossil on our behalf. One of her students, Ross Damiani, then made a detailed study of the specimen. This resulted in a significant scientific paper on the skull being published in *Alcheringa*, Australia's palaeontological journal. The specimen, identified as *Watsonisuchus aliciae*, is now on display in the Museum's 'Diamonds to Dinosaurs' Gallery.

In order to purchase such specimens from dealers, we museum curators often have to make special pleas for additional funds, or try to secure funds out of windfall accounts. For example, in 1995 the Museum made some profits through the 'Great Russian Dinosaurs' Exhibition and I was able to persuade our director that we should put back something into our collection from the profits. He agreed, and we bought a rare Russian fossil amphibian skull, *Thoosuchus jacovlevi*, from Tom for AU$1500. At the time there was some concern about fossils stolen from Russian museums being sold on the open market (see Chapters 5 and 6), so before buying it I made sure it was strictly legitimate. The specimen was examined by Dr Warren and the matter discussed with Dr Yuri Gubin of the Moscow Palaeontological Institute, who happened to be visiting Australia at the time. Both declared that, as far as they knew, the specimen we were buying had probably come from a private collection and was not one of the missing ones, which had Russian museum numbers written on them. Even if these numbers had been painted over or ground away, there would have been telltale traces left on the bone.

I asked Tom about his Chinese dinosaur eggs.

'How do you get them?'

'Dealers from China sell them at the international trade fairs—Tucson, Munich or Tokyo—and they arrange to ship them anywhere in the world for you,' he answered.

'But do they come with legal export papers from China?'

'Yes,' he replied. 'If you want them they can always supply you with papers written in Chinese.'

Most of us in the world of palaeontology know that any Chinese dinosaur eggs or fossils that turn up outside China have been illegally smuggled out of the country. The situation here is that the act of buying the eggs in the USA, Japan or Germany is not breaking any Australian law. If, however, the Chinese government were to issue a diplomatic request to a country where the eggs are being offered for sale, stating that they are of significant heritage value to China and they want them back, then that country would be under considerable pressure to seize the illegal goods and return them to China. More so if both countries are signatories to the UNESCO agreements about return of cultural heritage, as are Australia and China. For some reason this has not yet happened, although Chinese scientists have in the past requested the return of material from major museums, but these requests have not come from a high diplomatic level.

Tom and I moved on to discuss Australian fossil collectors who spend big bucks. Tom estimated that he knew of about 50 collectors who would spend up to AU$50 000 each year on fossils and minerals. This estimate was way above my own guess but, then again, Tom knows the market and I don't. Still, it amounts to total sales at the top end of around AU$2.5 million. In February 1992, the *Bulletin* magazine quoted Willyama Minerals dealer Kevin Davy as saying that the estimated 'total legitimate fossil sales in Australia would not exceed $100 000 annually'—this new estimate shows the extent of the industry's rapid growth in Australia (Bunk 1992). Tom further emphasised that as most fossil sales are in the lower end of the market, one can only estimate a much higher turnover for total fossil-related sales in Australia and, of course, for Australian material sold at international trade fairs. The bulk of the trade in Australia, though, would more than likely be in foreign fossils (from Morocco, the USA and China), rather than local specimens, as these constitute most of the product currently available on the fossil trade market.

I asked Tom about the stolen Broome dinosaur prints,

and where he thought they may have ended up. He agreed with us that such a fossil would not have much actual commercial value, as dinosaur footprints quite often appear on the international market and sell for small amounts, as little as a few hundred dollars. Tom thought that the footprints probably would be still in the country, either in somebody's shed or being used as a doorstop. In Tom's view they are a very specialist item, not commonly traded, except to a collector who particularly likes dinosaur footprints.

How could such prints be sold overseas? Tom's answer was, 'It's possible, but unlikely.' Simply take them to one of the big international trade fairs and keep them 'out the back'. By letting the right people know you had them, contacts could be made to put you in touch with a suitable collector. As most of the world's major natural history museums had been notified through widespread media coverage that the Broome dinosaur prints were hot, none of them would be interested in buying them. Tom's advice was to go and make further enquiries at a large fossil trade fair, such as Denver or Tucson.

We thanked Tom for his time, and left for Melbourne airport. Next stop, Sydney.

Opalised fossils

Monday 3 September. I go into the Australian Museum to talk with my old friend, Dr Alex Ritchie. Alex worked in the Australian Museum for over 25 years and since his retirement some years ago has been working there voluntarily. He has had many dealings with fossil sellers, and was happy to talk to me about them over a beer at the nearby Museum Hotel.

Alex told me the story of how Australia's first Mesozoic mammal—a mammal from the age of dinosaurs—was acquired back in 1984. He spotted it in a collection of opalised fossils, owned by the Galman Brothers from

Lightning Ridge in New South Wales. His eyes nearly popped out of his head when he saw the little jaw, completely taken over by opal, but still showing the incredible details of the three complex teeth that told Alex it was definitely a mammal jaw. It was later named as *Steropodon,* meaning 'lightning tooth'. It pushed back the presence of mammals in Australia from about 30 million years to nearly 120 million years. The Australian Museum, after much persuasion by Alex, eventually found a sponsor and acquired the entire collection, just to get that jaw. They paid AU$85 000 for the collection, the most any museum in Australia had ever paid for fossils purchased through a dealer.

The story of Eric, the opalised pliosaur, is another great case. It is mainly down to Alex that Eric is safely within the collections of the Australian Museum today. Eric was found in 1987 by Joe Vida, an opal miner at Coober Pedy. The opal mining machinery crushed the complete articulated skeleton into hundreds of fragments. Local entrepreneur Syd Londish purchased it for around AU$125 000 and had the idea that he would have it prepared by the Australian Museum. According to Alex, Londish had every intention of donating it to the Australian Museum under the Tax Incentives for the Arts scheme. Paul Willis, then a doctorate student, but later to become a well-known 'Quantum' presenter for the ABC, spent some 450 hours of meticulous labour preparing the bones out of the rocks and joining all the broken fragments together.

The end result was extraordinary—a virtually complete articulated skeleton of a small seal-like reptile, a pliosaur, which experts placed in the genus *Leptocleidus*. Almost as soon as Paul Willis had finished preparing it, Syd Londish went bankrupt, so the bank promptly seized his assets, including the fossil specimen, which Paul had christened 'Eric' after the Monty Python song about 'Eric the half a bee' (as Eric was almost, but not quite, complete). Eric was now going up for auction to the highest bidder, and Australian palaeontologists had to try to scotch plans for the fossil to end up overseas.

Dr Rupert Wild of Stuttgart Museum told me about one such plan. A major German industrial company offered Eric to Stuttgart, probably expecting to buy the fossil outright then donate it to the museum. Wild refused the offer, well aware that Australia's heritage laws would never allow such a precious specimen to be exported. (Especially in view of previous cases where Australian opalised plesiosaur skeletons were sold to overseas museums, prior to the introduction of the *Protection of Moveable Cultural Heritage Act* and, I might add, in complete violation of the *Australian Customs Act* of 1909. One such specimen now resides in a natural history museum at St Paul, Minnesota.)

Alex wanted Eric for the Australian Museum. He had known Karina Kelly, presenter of the TV science show 'Quantum', for many years; apparently she was on board the same ship on which he came out to Australia. They had been in contact over the years through various ABC stories and so on. It was now time to call in a few big favours. Alex asked Karina if she could launch an appeal to the Australian public to donate money to save Eric. The appeal was launched and money flooded in. In addition to a large donation from Australian hat manufacturer Akubra, the Australian Museum received over AU$450 000. The day of the auction was only a few weeks away, but Alex was now confident that the museum could put in a reasonably significant bid to secure the specimen. The auction was by silent bid. The Australian Museum put in a bid of AU$330 000 and it was accepted. They had won. Eric was now theirs forever and they had a healthy balance of funds to be used for other fossil acquisitions. As a 'thank you' to sponsors and the members of the Australian public who bought Eric, the specimen went on an Australia-wide tour so that as many as possible of those who donated towards saving Eric would have the chance to get a close-up look at the actual specimen. I remember delicately placing Eric in his custom-made glass case when he toured around Western Australia, from Perth to Geraldton, Albany and Kalgoorlie.

'Where is Eric these days?' I asked Alex.

'Come along, and I'll show you. It's only a short walk downtown.'

We walked down to the centre of Sydney's business district. Eric now lives in an underground opal shop, Cody Opals at 176 Pitt Street, as the Australian Museum had made a deal to rent the specimen out for dollar returns, much needed by the museum's administration. It seems that despite all the wrangles over the ownership and safe conservation of the specimen, the bottom line is that even the museum sees Eric as a dollar commodity. Of course, it also gives the Australian Museum a strong public presence in central Sydney, and all the visitors who marvel at the beauty of Eric and the other wonderful opalised fossils in the shop can be lured to visit the Australian Museum to see more fossils. Unfortunately, the specimen remains undescribed, as detailed scientific analysis has not yet been carried out and it is not easily accessible for those wishing to study it.

We travelled on to meet Warren Somerville, a slight man in his mid-fifties with an intense passion for fossils. Warren is a farmer from Orange, New South Wales, who has gradually amassed Australia's largest private fossil collection. Warren has specimens from all around the world. Today his collections are valued at around AU$7–14 million (depends on the market), and the Australian Museum in Sydney has recently taken steps to make sure they will acquire the collection when Somerville dies.

I first met Warren when we were both called as expert witnesses in the trial of the fossil dealers who were caught exporting material without a permit in 1991. I remember taking him to see one of the specimens we had recently bought, a large slab of fossil fishes from the Green River Formation, Wyoming, for which we paid the princely sum of AU$2800. At the time I had to argue to our administration that it was a good buy and would work well in our new fossil and minerals gallery (which was then still on the drawing board). I showed the specimen to Warren and asked him, in his official capacity as an expert valuer, how much he thought the slab was worth as he had several similar specimens in his collection. I was delighted with his answer—we had got a bargain basement price on the slab. This was of considerable significance to the court case, as one of the defence's arguments against me was that professional palaeontologists are not capable of accurately estimating the market value of fossil specimens (at this time there was a AU$1000 price cap on specimens requiring an export permit).

I next worked with Warren some years later, when the Western Australian Museum needed two expert valuations of the worth of a private fossil collection that was possibly going to be donated to us. Warren spent about three days working his way through the collection, valuing every specimen or specimen lot according to current estimated market values. He did a splendid job. We were able to convince the owner of the collection that he would be well reimbursed with tax credits for donating the material to us, and that's exactly what transpired.

Warren's own collection includes a cast of Stan, a famous *Tyrannosaurus rex* skeleton fully thirteen metres long, as well as a complete plesiosaur from Morocco (possibly an undescribed new genus, which makes it very interesting scientifically). He has scores of complete fossil fish, dinosaur eggs, myriad ammonites, trilobites and nautiloids, and many types of fossil plants. His material is enough for a whole museum, and a new museum, designed around his collections, is scheduled to open in Bathurst in early 2003.

We were particularly interested in interviewing Warren because he was one of the private fossil collectors to whom Latham had offered the dinosaur footprints stolen in 1998. Warren remembers that he was asking a ridiculous amount, AU$250 000, for the stolen fossils. He declined, not only because all the media coverage about the previous fossil theft from Broome made it obvious that this specimen was clearly illegal, but also because the actual specimen wasn't that special, certainly not worth anywhere near the asking price. (Our later investigations at the Tucson Fossil and Mineral Show would show that good-quality dinosaur footprints can be bought for between US$250–$1000, perhaps slightly more if two or more prints are preserved in the same slab of rock.)

Warren was also asked where he thought the missing dinosaur prints might have ended up. He agreed with Tom Kapitany that an overseas collector with a penchant for dinosaur footprints would be the most likely destination. We thanked him and made tracks for the airport hotel.

That night, before we left for London, Steve and I discussed the legality of fossil finds in Australia and I gave him a brief overview of the rights of those who find fossils on leased or government-owned lands. I was involved in one such case in 1993, that of the giant elephant bird egg.

The case of the giant elephant bird egg

In January 1993, nine-year-old Jamie Andrich was playing in the high sand dunes around the beach just north of Cervantes, Western Australia, when he made a remarkable find. Poking out of the sand was a giant white egg. A few weeks later, he and his parents brought the egg in to the Western Australian Museum and I found it sitting in a cardboard box on my desk when I came back from lunch. I immediately recognised it as an egg of the Madagascan elephant bird, *Aepyornis maximus*, not just by virtue of its huge size (31.7 cm long) but because it wasn't the first time a giant elephant bird's egg had been found in a Western Australian sand dune. A similar egg was found in 1930 near the mouth of the Scott River, Augusta, by a Vic Roberts. This egg was already in the Museum's vaults. The next day, after I had identified the egg and guessed it to be around 2000 years old, the Andrichs came in to the Museum and claimed it back.

At the time of the discovery, Jamie's family immediately assumed that it was a simple case of 'finders keepers'. They were told the egg could be worth a lot of money and were prepared to sell it to the highest bidder. They employed a professional auctioneer to handle the sale of the egg and to generate publicity about it. They invited me to see the site it came from, so I visited the dunes, took some photos and did some measurements on its stratigraphic position within the dune system. It was found about 300 metres inland, only about two metres or so above current sea levels. The egg's auction generated an enormous amount of media interest. Claims from the auctioneer that it was worth AU$150 000 came in, as well as anonymous bids from a Japanese company. While this was going on, I was researching the egg and its value, based on similar finds and market value at recent auctions. My estimate was that the egg was worth about AU$5000–$10 000, because that was the going price for an *Aepyornis* egg at auctions in England. It was at this point that the vendors took the egg to be examined by experts in Melbourne and Sydney and get a small part of it radiocarbon-dated. It was soon confirmed that my identification of *Aepyornis maximus* was correct, and that my age estimate of 2000 years (a lucky guess on my part) was close to the carbon age of 2000 +/– 75 years.

The ownership of the egg, however, still had not been adequately resolved. The egg had been found on Crown Reserve Land, not private land, so we argued that by all rights it should belong to the Crown (namely the State of Western Australia). Crown lawyers began to consider the case in detail. About six months later, after much more media publicity, and numerous attempts to sell it overseas being thwarted, we were able to ensure that the specimen would never be exported (by export permits being refused), after the Crown legal department handed down its findings that the egg was indeed Crown property under the *Crown Lands Act*.

The Andrich family was infuriated by this and promptly went back to Cervantes and re-buried the egg in the sand dunes. On advice from their lawyers, however, they soon retrieved the egg and struck a deal with the government of Western Australia: the egg would be publicly displayed at a prominent city bank, and an appeal launched to raise money for the family. The government gave them an ex-gratia payment of AU$25 000, a similar amount to that given in the past to finders of shipwrecks or other archae-ological treasures. The unsuccessful public appeal raised only $500 in extra funds.

Today the egg sits in the Western Australian Museum's 'Diamonds to Dinosaurs' display. A scientific paper (Long *et al.* 1998) argues that both it and the Scott River egg could have floated as addled eggs across the Indian Ocean from Madagascar before being buried in the dunes. The case also resulted in a legal paper, published in the *Australian Property Law Journal*, which reviewed ownership of title by finding (Tooher 1998). An interesting aspect of the case was that it made the government of Western Australia aware of the lack of suitable fossil legislation in the State, and the need to protect sites and to clarify the position of owner-ship of any future finds—not just those found on Crown land. In this case it was pure luck that the specimen was found on Crown land. Had it been found on private property the State would have had no legitimate claim to it. The case of the giant elephant bird egg established a useful precedent about ownership of fossils under the *Crown Lands Act*.

I must admit that I felt very uneasy about such a rare and delicate specimen being in private hands. It could easily have been accidentally dropped or, had ownership reverted to the family, smuggled overseas and sold to a private dealer. Even more worrying was the possibility that someone might have come up with the idea that it would be worth more money if sold piecemeal, as 1 cm-long pieces in a necklace (for example), then the egg could have been smashed into a thousand pieces and sold in that fashion.

It is a scandal that, at present, we have no laws in Australia to protect or conserve adequately scientific heritage items which are held in private collections. The case of the giant elephant bird egg was the bane of my life at the museum for more than three years. The ensuing media and public enquiries, from all around the world, were numerous and time-consuming. In the end I wrote a formal publication on the egg in order to make the information on the case available to the public. I still squirm when I think of some of the headlines at the time, when the Minister of Arts' decisions were being openly criticised by the media, who tended to side with the Andrich family. 'Minister with poached egg on face' was just one of them. To my mind, however, the best line belongs to my colleague Dr Alex Bevan, who had to deal with the handover of the egg to the museum while I was away overseas. He sent a fax about the Cervantes egg business to a colleague in France saying 'Un oeuff is un oeuff [sic]!'.

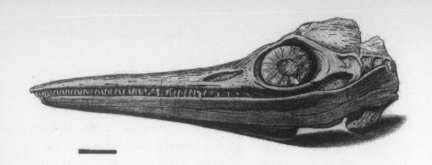

London Calling

5

'Have ye sid my animal sir,' said the fossilist Jonas Wishcombe of Charmouth as I called at his house in August to enquire if he had anything worth buying:— 'I should like vor yer honor sir to see un.'

My heart leaped to my lips—'Animal! animal! where!'

'Can't be sid to day sir—the tide is in.'

'What? Nonsense—I must instantly—come, come along.'

'Can't see 'un now yer honor, the tides rolling atop o'un fifty feet high.'

'In marl or stone?'

'Why in beautiful ma-arl,'—and—Washed to death—and I threw myself in despair upon a chair. How often have I reflected upon the very Bedlam impetuosity of my passions at that moment:—the chaffed sea rolling over an *Ichthyosaurus* and remorselessly tearing it into a thousand atoms—a superb skeleton of untold value triturated to sand by a million pebbles . . .

Sketch of the skull of an ichthyosaur sold by Mary Anning to the British Museum of Natural History. (Bar scale is 10 cm)

(Thomas Hawkins, *Memoirs of Ichthyosauri and Plesiosauri*, 1834).

No wonder Thomas Hawkins was plunged into despair to hear about the perfect *Ichthyosaurus* skeleton being eroded by the Charmouth seas. Thomas Hawkins, of Glastonbury, was one of the most energetic English fossil collectors of the 1830s. In 1834 he sold his first private collection of marine reptiles and other fossils to the British Museum of Natural History for the enormous sum of £3000. He would pay quarrymen to keep any fossil bones they came across from the Jurassic-age quarries near Edgarley in Somerset, England. Fossil collecting for the purpose of selling the specimens to learned institutions was a respectable profession.

One of the best known of the early professional fossil dealers was Mary Anning from Lyme Regis, the daughter of Richard Anning, a cabinet maker and occasional fossil seller. Richard died in 1810, leaving his family almost destitute. He had, however, passed on his fossil-hunting skills to his wife and children, which would prove fortuitous for the fledgling field of palaeontology. From the late 1700s through the early 1800s, Mary and her family made a living collecting and selling numerous marine fossils from the southern coast of England.

Most of the material was from the Jurassic period, such as ammonites or shells, but she and her brother Joseph collected and sold the first identified skeleton of an ichthyosaur, a giant marine reptile, to Henry Host Henley of Sandringham in Norfolk for £23. The specimen was seventeen feet long with a four-foot skull. Found in late 1810, by November of 1812 the family had raised sufficient funds to hire quarrymen to excavate it out of the cliffs between Charmouth and Lyme Regis. Henley deposited the specimen in William Bullock's London Museum of Natural History in Piccadilly, and the specimen was described by Everard Home in 1814. Both Mary Anning and her mother remained active in the fossil business for the next 30 years. Mary Anning is reputed to have inspired the well-known tongue twister, 'She sells sea shells by the sea shore'.

Another keen amateur, Colonel Thomas James Birch, of Lincolnshire, would spend his winters in the West Country acquiring fossils, purchasing them from professional collectors at Lyme Regis and other sites. Birch auctioned his collection in 1820 in order to raise money on behalf of the Anning family, as his stay with them in 1819 had shown him that they were experiencing major financial difficulties. It seems that the fossil-selling business in those days was just as much at the mercy of Lady Luck and the highs and lows of market demand as it is today.

Searching for clues in downtown London

Friday 7 September, 2001. Chasing up one of the brumours we were told, we go down to central London to look for a shop rumoured to sell curios and antiquities from the collection of Lord Alistair McAlpine—'Erebus', named after the mysterious and isolated volcano in Antarctica. We think we might find some interesting fossils there, if it's still in existence.

We found the place, going by an address we had been given, but there is no obvious entrance to the building, nor any signs of the shop's existence. We were told that the shop didn't have a frontage, but was inside a building on Cork Street. Steve and I decided to ask some of the local antique shop dealers if they knew of the shop. The first gallery we went into was the Gordon Reece Gallery. After having a look at the superb range of antiques in the shop, we made light conversation with the shop's owner, Gordon Reece. Steve asked Mr Reece if Lord McAlpine had an antiquities shop near here.

'No, he has no shop around here. He's a Lord. He certainly wouldn't be running a shop these days.'

That afternoon we visited art and antique galleries, in

Soho, where we discussed Australian art with the owners and marvelled at their collections, much of it from the Kimberley district. There were no Australian fossils on display in the galleries, only Aboriginal art that would have been obtained through proper export permits.

At this stage we had no hard evidence for the dinosaur footprints ever having come to London, despite the various brumours we heard, so we decided to leave the antique galleries of Soho.

Saturday 8 September. That morning, we stop for a coffee and discuss our options.

Suddenly Steve went quiet, and he motioned to us to lower our voices. His eyes darted around the crowd. Apparently we were being watched by a man drinking coffee a few tables over from us. After our experience in Broome, I took this information very seriously. We weren't on home ground, and had to be extremely careful. Steve decided to test his theory about us being followed.

'Come on John,' he said casually. 'Let's go find some *real* coffee around here. I saw a Starbucks a few blocks away.' The film crew decided to take the equipment back to the van, and we told them we would all meet up in fifteen minutes or so back at the car park. Steve and I strolled slowly away from the London Eye, taking a circuitous route around two blocks to the Starbucks. I should point out here that Steve is a real devotee of Starbucks coffee, and would normally take the most direct route, at a pretty brisk pace, to get his Cafe Grande Americaine with five shots of espresso. This time, however, we ambled down a busy street with mirrored glass windows on both sides. Steve watched the reflections across the street carefully. We entered the Starbucks, ordered our coffee and sat down near the window.

'Yep, they were on our tail all right,' he said calmly.

'We'll have to shake 'em off going back to the van.'

I suddenly felt a little nervous about the whole thing. After all, I'm just a regular palaeontologist who came along to identify fossils along the way, and to make the positive identification on the stolen Broome prints if they turned up somewhere. Now I'm holed up in a downtown London Starbucks coffee house, with a top US cop who tells me we are being followed. All the cockney London cop shows I'd ever watched went racing though my head, along with scenes from *Lock, Stock and Two Smoking Barrels* and *Snatch*, about the bad guys being ruthless in taking out their opposition. I decided I would stick close to Steve at all costs and say nothing.

'Let's move,' Steve said after we had downed the hot coffee. I was now high on a triple shot of caffeine, and ready to race Steve to the van. We walked quickly around the block, backtracked half a block on the other side of the street, then surreptitiously cut through a narrow parkway to emerge not far from the London Eye. Crowds were gathering around the tourist area, a good thing for us, so we walked briskly through the swarm of people, down the street and into the car park.

'Quick, get outta here!' Steve urged Alan. The van sped away into the narrow London streets, and an hour or so later we were back in our hotel.

We decided we would play it very low key in London from then on. Alan had realised that our leads didn't seem to be going anywhere, yet for some reason we were being carefully watched. On Steve's advice, we decided that we would rest the following day and take stock of all that had transpired so far. That night we were all a bit overcome by jet lag and nerves. None of us went out to have dinner or a beer; we simply ordered from room service.

Sunday 9 September. Steve gets up early and goes for a long walk through the streets of London, by himself.

On returning to the hotel early that afternoon, he told us that he had been followed all day by two young ladies. He had tried walking down various laneways and changing directions at random, but all his attempts to shake them off had failed. In the end he used the direct approach. He led them to a downstairs McDonalds restaurant where there was no other exit. As they came down the stairs, he turned around and walked towards them, confronting them face-to-face halfway up the stairs.

'Why are you following me?' he asked them directly.

'Eh, what do you mean . . .?' one said nervously, her eyes looking everywhere but at Steve. 'We weren't following you.'

'Yes you were,' Steve replied curtly. 'Now leave me alone or I'll have to call in the police.'

He walked quickly away, watching out of the corner of his eye to make sure they remained in the restaurant. He jumped into a cab and came straight back to us at the hotel. To this day, the reasons why Steve and I were tailed around London remain a mystery to us. We have no idea why we should attract such interest, but we took it seriously just the same.

Some time later we finally made contact with Lord McAlpine, who agreed readily to meet with us and talk about the case. Lord McAlpine greeted us cordially and we began by chatting about Broome and his early memories of Australia. On raising the subject of the missing dinosaur footprints, he became agitated, as he had read in a newspaper that his name had been mentioned in connection with Michael Latham's recent trial. He said that his connection with Travelstead had been wrongly reported; he had never used Travelstead as his 'agent' for buying Aboriginal art and artefacts.

'There is no evidence that I was interested in the footprints, and I wouldn't have been interested, full stop,' he told me. Through his friendship with Joseph Roe's father, Paddy Roe, Lord McAlpine was all too aware of the spiritual value of the missing footprints.

'Paddy was very kind to me and he explained to me the importance of the landscape and the religious ideas,' he began. 'I could understand that a piece of rock could be sacred; after all we've crowned monarchs in England on a piece of rock—the Stone of Scone—for the last 700 years.

> The problem about thieves is that they often steal things that aren't valuable because they think they're valuable and the real sadness is when they can't sell it [sic], for a lot of money, they throw it into the sea. I suspect that's what happened to these.
>
> I just think the whole thing's a tragedy more in its stupidity than its evil intention . . .

We thanked Lord McAlpine for his time and returned to our hotel. It seemed that the stegosaur prints may have been too hot for any collector to handle. If Lord McAlpine was correct, then instead of a 'stolen to order' conspiracy, the theft may well have been just a stupid mistake. Further dialogue with Michael Latham in Broome might be our only chance to clear up the confusion and Steve decided to give him a call.

Steve made the call to Broome Prison, and was lucky enough to get straight through to Michael Latham. At first Latham was wary about talking to Steve, but eventually he relaxed and spoke with Steve for about fifteen minutes. Although Latham did not give much away he hinted that his brother Dennis had also been involved in the theft of dinosaur footprints. Steve's gut feeling after the call was that Latham had told him mixed truths; nonetheless, he thought that we should keep in touch with Latham, as he would more than likely eventually give away some clues.

That afternoon, while Steve was walking around London, I visited the Orpington home of my good friend and palaeontological colleague David Ward. David is a world-renowned expert in fossil sharks, whose work often takes him around the world. When I told him that we were investigating the international fossil trade in our search for the missing footprints, he got very excited.

'Have you heard the tale of the stolen *Helicoprion*?' he asked eagerly.

'No,' I replied. (*Helicoprion* was an amazing fossil shark of the Permian Period whose lower jaw formed a bizarre coil of jagged teeth.)

'Good, get out your notebook. I was personally involved in the case,' he said, 'so I can tell you the whole story as it really happened.'

The case of the stolen tooth whorl

Helicoprion bessonowi is one of the world's great enigmas. When it was found in the Russian Ural Mountains in 1899 it aroused great curiosity in the world of natural history. It was a single, curled row of shark's teeth, some 125 in all, which looked more like an ammonite than a shark's jaw. Each tooth was serrated and overlapped the front of its predecessor. *Helicoprion* whorls indicate that the whole shark must have been a monster, perhaps 8–12 metres in length. We can only hypothesise that the lower jaw tooth whorls hung out of the shark's mouth and were used to snag fish as the huge predator charged into densely-packed schools of ray-finned fishes. The holotype (the original specimen from which the scientific description is made) of *Helicoprion bessonowi* was described by Karpinsky and then deposited in the Natural History Museum in St Petersburg, Russia. It remained there for all to examine and study, in the locked vaults, for almost a century, until it was brazenly stolen.

In November 1998, a good friend of David's, Vinny, told him he'd just been offered a specimen of *Helicoprion* by a reputable Russian dealer. Vinny had been told that the specimen came with all the requisite paperwork to allow its export. Russian fossils are rare and very marketable, so a price of US$2500 was agreed, based on a photograph of the specimen, which was then dispatched. Pondering his purchase, Vinny turned to David's famous guidebook, *Fossils* (Walker & Ward 1992, p. 198), to find out more about it. The illustration in the book seemed very similar to the photo of the Russian specimen he had just bought but, as he rotated the photo, it changed from being similar to being identical—it was the same fossil! A somewhat perplexed Vinny immediately rang David and asked him about the specimen. Almost all the photographs in David's book were of specimens in the Natural History Museum in London, so a very worried David lost no time in visiting the Museum where, to his great relief, he found the *Helicoprion* specimen, safe in its drawer. To his surprise, however, he found that the specimen was, and always had been, a plaster replica. A note on the label said that the original was in the CNIGR Museum in St Petersburg. David emailed a Russian friend, Sasha Averianov, a palaeontologist at St Petersburg University. Sasha replied promptly: 'Thanks for your message. The holotype of *Helicoprion bessonowi* should be in the CNIGR Museum (St Petersburg) collection. I saw it there a year ago—I will check again. The museum does not sell fossils'.

On hearing this, and thinking of the previous fossil theft from the Russian Palaeontological Institute in Moscow, David phoned his friend Sasha Ivanov at St Petersburg Museum, and asked him to check on the holotype of *Helicoprion bessonowi*.

'It's gone!' came the reply.

Two weeks passed, then the following announcement was posted on the Web.

To everybody whom it can concern. From St Petersburg (Russia) Museum of Geological Research (name Chernishova) was stolen Very important sample of *Helicoprion bessonowi* 'holotipe' sample.

(Photo available) If some one now anything ore here ore something Please help us to return it to museum ore to us personal. Address of museum Central Scientific Research Museum name Chernishova 199178 Sredniy prospect Vasilevskiy ostrov 74 St Petersburg Russia Please help museum if you now something, confidential is guaranteed. Thank for attention
Arkadiy
PPL. St Petersburg Russia

A precious specimen which was kept locked up in the type room vault somehow had been stolen. David phoned Vinny in Florida, and told him to cancel his cheque. Luckily, it hadn't yet been presented. David then phoned his colleague, Gordon Hubbell, another sharks' teeth expert, and asked him to examine Vinny's purchase when it arrived and confirm that it was the missing holotype. Sure enough, the specimen arrived in Florida, having been shipped from Finland. Gordon confirmed that it was indeed the original holotype. David rang the Director of the St Petersburg Museum to let him know that the holotype was safe and in good condition. He vowed to return it personally to the Museum. (By chance, David had already arranged to visit St Petersburg in a few months, to photograph specimens from the very same collection.)

It didn't take much persuading for Vinny to agree to return the specimen: as soon as it was confirmed that it was 'hot' he couldn't get rid of it fast enough. He agreed to ship it to a friend of David's who lived in California, who then met David in Paris at New Year. The *Helicoprion* spent the late winter and early spring in a hatbox in the corner of David's study.

When the time came for David to return the *Helicoprion* to St Petersburg, the idea of carrying a stolen fossil back into Russia seemed increasingly unwise. Should he declare it or keep quiet? On the advice of his colleague Dr David Unwin, who had spent much time in Russia, he took some precautions.

First, he wrote to the Director of the Museum and requested a formal six-month loan of the holotype specimen, with permission to be written in both Russian and English. Director Karpunin was most cooperative and furnished a suitably impressive document with the obligatory signature over an official museum stamp. Next, David asked a translator to draw up a document, in Russian, confirming that he was returning the fossil at the expiry of the loan agreement.

The big day came and David set off for St Petersburg, with the *Helicoprion*, in his briefcase, as hand baggage. Despite being one of the first off the plane, David found himself near the end of the line as he approached the customs desk. Almost immediately he was singled out and taken to a private room for further questioning. He started to sweat, thinking that he might have been set up by someone with a vested interest in the specimen. Perhaps the dealer who had had it stolen. Some of these dealers, he had been warned, had powerful underground connections. The customs official asked him why he had no luggage, and told him that he would have to fill out another form. She didn't even ask to see inside the briefcase containing the *Helicoprion* specimen. After he had filled out the correct form, she waved him through the line.

Outside, the Director of the St Petersburg Museum and David's friend Sasha were waiting for him. When they arrived at the museum David officially handed over the specimen. There was a ceremonial unwrapping of the specimen before it was gently placed back in its drawer in the vault, to which new locks had been fitted. The *Helicoprion*'s return was toasted with special German brandy and celebrated with coffee and cakes. For his dedicated work in returning the priceless specimen, David received a certificate and medal of honour from the board of directors of the museum.

We still don't know who orchestrated the theft, but they must have relied on help from within the museum, someone who had access to the keys of the vault. At the same time as news of the theft broke to the world, a German dealer, Joachim Wördemann, was arrested in St Petersburg with a Mercedes van full of Russian fossils (Abbott 1999; see Chapter 6). He was questioned over the *Helicoprion* theft, but claimed he knew nothing about it.

David and his friends had T-shirts made up to commemorate the event, and he gave me one (the T-shirts are available from Paleoworks). On the front is a colour photograph of the specimen with *Helicoprion* written underneath. The legend on the back of the shirt reads:

Helicoprion
Grand World Tour.
Permian to 1899 Ural Mountains, Krasnoutismk, Diva-Cora, Russia.
1899 to 1998 St Petersburg, Russia.
November 1998 Finland–Florida, USA.
December 1998 California, USA.
December 1998 Paris, France.
January 1999, London, UK.
May 1999, St Petersburg, CNIGR Museum, Russia.

Let's all hope it stays there!

The following case is one example which involved specimens taken from Britain and sold in Germany. The specimens were fossil fish and invertebrates which have been found at only one locality in the world, so it was relatively easy to trace their provenance.

A case of stolen Scottish fossils

31 October 2000. Silurian fish and invertebrate fossils from Scotland are stolen from a protected site, and sold to a museum in Germany. Scottish Natural Heritage are alerted in February 2001 after an academic reports seeing the 430-million-year-old remains in a museum in Berlin.

A team of Scottish Natural Heritage geologists decided to visit the Humboldt University Museum in Berlin, to discuss the possible return of the rare fossils. *Jamoytius kerwoodi*, an anaspid (one of the eel-like jawless fishes), has only been found on a river bank in Lanarkshire, which has been designated a Site of Special Scientific Interest. This site may have been raided several times by a foreign collector who knew what he was looking for.

Several well-preserved fossils of *Ainiktozoon*, a proto-chordate (related to the first creatures with a backbone), were destroyed in the collecting raid. The well-preserved, complete fossils of *Jamoytius,* each about 16 cm long, are worth about £10 000. There are believed to be only about 150 *Ainiktozoon loganese* fossils in the world, all of which come from the one site in Scotland. Colin MacFadyen, the geologist leading the delegation, accepted that the German museum had bought the fossils in good faith, but he thought it was important to establish that the remains had been obtained without permission of the landowner and without a licence.

As our plane departed Heathrow, en route for Germany, I wondered just what we were getting ourselves involved in. Would Germany provide us with the clues we needed to crack the case, or would it prove to be a dangerous place for people like us snooping around asking questions about stolen fossils? I had read a lot of articles about stolen or illegally exported fossils turning up in German shops, or being confiscated from German dealers, so it seemed to me to be a good place for us to try to trace the path of a valuable stolen fossil, or at least to uncover information about possible end destinations of black-market fossils.

Undercover
in Hamburg

6

Under the UNESCO convention, state parties must act to prevent museums from acquiring illegally exported cultural property (which includes fossil materials). Germany is not at present a member of UNESCO; the UK has only recently announced its accession.

Sketch of the famous Maastricht mosasaur stolen by Napoleon's army. The lower jaw is nearly a metre long.

An early case of fossil theft

As our plane landed in Hamburg my mind drifted back to Europe's history as a centre for palaeontological studies, and its importance in the history of fossil trading, fossil theft and even fossil fakery. These two cases are good examples.

Mosasaurs were one of the earliest groups of fossil marine reptiles ever to be discovered, the first found about 40 years before the first dinosaur was discovered. This was the famous Maastricht mosasaur, uncovered in 1780 near the town of Maastricht, in Holland. The specimen became the subject of a protracted legal battle between a Monsieur Hoffman, who oversaw the excavation, and the Canon Goddin, on whose land it was found. When the French army besieged Maastricht in 1795, the house where the mosasaur specimen was kept escaped the bombardment. Suspecting that the French knew of the specimen's whereabouts, the owner hid it in a secret vault in the township. The French did know about the famous fossil and they wanted it so badly that they offered 600 bottles of wine for its recovery. Not surprisingly, it was found soon after, and today it resides in Paris, where it is on display in the Musée d'Histoire Naturelle in Paris. The specimen was eventually given the name *Mosasaurus*, meaning 'saurian from near the Meuse River'.

The Maastricht Museum of Natural History now has several good *Mosasaurus* specimens, and a new species of mosasaur has just been described, based on a complete skeleton some fourteen metres long, recently excavated from the same region.

Europe has a long history in the theft, trade and faking of fossils. Fifty years before the military theft of the Maastricht mosasaur was the remarkable case of Beringer's lying stones.

Beringer's lying stones

Dr Johann Bartholomew Adam Beringer was a German physician and virtuoso well known around Würzburg. Among his many interests was the study of 'oryctics', things dug out of the ground. He had collected many fossils himself and purchased many more from fossil sellers, but his collection contained nothing spectacular or different from his colleagues' collections. Then, in May 1725, three very different stones with strange markings were brought to his attention. One bore the image of the sun and its rays, the other two what looked like the outlines of worms. The finders claimed that the rocks had come from the slopes of Mt Eivelstadt, near Würzburg. Over the next few months more of the amazing stones turned up. Some had plants on them, others shapes of birds or various strange creatures, even astronomical hieroglyphs. Beringer became fascinated by them and began to collect them avidly, his mind intent on making a detailed study of them. He employed three youths to go and find as many stones as they could, as Mt Eivelstadt was soon identified as their sole locality.

Eventually Beringer had amassed over 200 of the stones, working tirelessly on his magnus opus, the *Lithographiae Wircebergensis*, which was published in 1726. Beringer saw the stones as containing significant fossils which had been formed under the sea. He called them 'iconoliths'. To modern palaeontologists who understand the fossilisation process, the preservation of soft-bodied animals such as slugs, spiders and worms as raised outlines in stone would immediately seem suspicious. But to Beringer, caught up too deeply in his own thesis and captivated by his large collection of fossil stones, everything seemed to fall into place.

Beringer had enemies, however, who had manufactured the stones to achieve his downfall. The main perpetrator was the University of Würzburg's Professor of Geology, Roderick Von Eckhart, but he was helped by the Privy

Councillor and Librarian to the Court of the Würzburg University. The stones were hand carved and placed on the hill to be found by Beringer's youths.

Soon after Beringer's book was published the truth behind the origin of the stones was revealed, showing his work to be a complete and utter waste of time. Beringer, racked by shame, tried desperately to buy back every copy of his precious monograph. Today, through his efforts, this book is one of the most infamous, and indeed the rarest, of all palaeontological works.

The German connection

We had come to Germany to meet up with a well-known fossil dealer in Hamburg who had been implicated in previous suspect fossil dealings, to see if he had any information that might help us.

One of the most notorious cases of fossil theft from museums in recent years is that at the Russian Palaeontological Institute in Moscow. The case was first reported internationally in the science journal *Nature* (Feder & Abbott 1994). Several well-preserved fossil amphibian skulls from Russia were stolen from the Institute in 1992. As the locked showcase from which the specimens were taken was intact, Russian palaeontologist Michael Shiskin suggested that someone from within the Institute must have been involved in the theft. Shishkin went on to report the theft in the international science journal *Lethaia*, so that any palaeontologists who saw such specimens for sale might recognise them and report them to the Russian museum.

A breakthrough in the case came shortly after it was reported, when German palaeontologist Dr Rupert Wild, of the State Museum for Natural History in Stuttgart, was examining a fossil amphibian skull from Russia that had been offered to him by a local fossil dealer. Wild noticed that it was very similar to one of the missing specimens. On closer examination he discovered that the skull, which his museum had purchased for DM1600, still bore traces of the Russian Palaeontological Institute registration number, visible only when the specimen was examined under ultraviolet light. The specimen had been sold to the museum by German fossil dealer Joachim Wördemann.

When Wördemann was questioned about the provenance of the stolen specimen he claimed to have bought it from a private collector (Feder & Abbott 1994). Consequently no charges were brought against him. Wördemann did the right thing and bought back the stolen skull, then returned it to the Russian Institute. It was personally collected by the Institute's Deputy Director, Igor Novikov, who hand-carried it back to Russia in September 1994.

In 1996 a series of further thefts from the Russian Institute was reported in *Nature* (Abbott 1996), this time involving rare Mongolian dinosaur specimens. Part of the skull (the upper and lower jaws) of the giant flesh-eating *Tarbosaurus bataar* (a close cousin of the famous *Tyrannosaurus rex*) was stolen, as well as skulls of the small horned dinosaurs *Protoceratops* and *Breviceratops*. All together, the fossils were valued at around US$11 000, but they were priceless to us scientists. The *Tarbosaurus* skull was the holotype specimen. An international working group of palaeontologists was set up, under the leadership of Dr David Unwin of Bristol University, to alert the palaeontological community about the stolen Russian fossils and to assist in their return. Despite repeated requests for help, however, administrators at the Russian Palaeontological Institute were not forthcoming in providing pictures of the stolen fossils. In January 1998, seven senior scientists from the Russian Institute sent a letter to the All Russian Palaeontological Society expressing their concerns about the security and long-term safety of the collections of the Institute. One of their concerns was that the Institute's Director, Alexei Rozanov, had tried to cut half the scientific staff positions (Abbott 1998).

Then, on 21 December 1999, Joachim Wördemann was stopped at the Russian border as he tried to drive a van full of Russian fossils, meteorites and minerals into Finland. Wördemann had licences to export most of the specimens, but he lacked the necessary permits for some of the goods. This was the point at which authorities started to become suspicious. Most of these licences were countersigned by directors of the Russian Palaeontological Institute. Under Russian heritage laws, fossils that are scientifically important are never allowed export licences. The previous theft from the Institute, and the involvement of the same German dealer, was simply too much of a coincidence. The timing was perfect to focus world attention on fossil thefts from museums, because the holotype specimen of the shark *Helicoprion bessonowi* had just been stolen from the museum in St Petersburg (see Chapter 5).

Wördemann visited Russia at least twice a year and was regularly seen at the Institute. Another article in *Nature* put an estimated value of millions of US dollars on the haul in his van (Abbott 1999), which included specimens that would certainly be considered of scientific importance, such as complete baby mammoth skulls (showing the developing teeth) from the Southern Urals and many well-preserved ammonites and trilobites, some of them possibly new species still to be formally described. Wördemann's arrest should have resulted in either a severe fine or a jail sentence, but apparently he was let off without being involved in a major court case, although he did lose his new Mercedes van to the customs office at St Petersburg. He operates a thriving business in Russian fossils to this day, but doesn't work from a shop. Instead, most of his specimens are sold at international trade fairs or through private channels, mainly as cash deals.

In order to find out as much information as we could about the shadier side of fossil dealing without raising suspicion,

Steve suggested we go undercover, and pretend to be businessmen, without any real knowledge of fossils, but interested in buying some prize pieces. Initially, I was a bit anxious about this approach but, on thinking about it for a while, I realised that I work in a museum, and our collections are sacrosanct. If anyone stole fossils from the Museum, especially type material, I would be livid and would want them caught. If we could catch one of the dealers unaware, we might unearth some clues to the whereabouts of our missing dinosaur tracks. Furthermore, we might find out something that could assist the Russian Palaeontological Institute, or even be a good thing for museum security worldwide. So, for these reasons, I agreed to go along with Steve's plan.

Before we left London for Germany, we contacted a well-known German palaeontologist and asked him for a list of prominent German fossil dealers. We also asked him which ones he thought might be involved in illegal activities, which might be big enough to deal with Australian fossils and so on. He faxed us the information a few days before we flew to Germany, so Steve and I had time to do some background research on the names on the list.

Tuesday 4 September. We arrive in Hamburg at about 9 am, then organise hire cars and arrange our accommodation.

From that moment on Steve and I went undercover. We booked into a different hotel from the film crew and assumed our new identities.

'Take all your name tags off your bags,' he told me as we left the airport. 'Now remove all identification from your wallet: credit cards, business cards and so on.'

We checked in to our hotel as Steve Richardson and John Lyndon, of RLA, Richardson Lyndon & Associates, business management consultants. I hoped they wouldn't want to see our passports. As we filled out the hotel reservation cards the receptionist smiled at us.

'How would you like to settle the account?' she asked in very good English.

'We'll pay cash,' said Steve, smiling at her.

'Fine,' she said, filing our cards away and giving us our room keys.

I breathed a sigh of relief as we walked upstairs.

'Now, that wasn't too difficult, Mr Lyndon, was it?' said Steve, chuckling to himself.

The plan was for us to take a walk around the docks area of Hamburg on camera, discussing ways and means that goods come in and go out of the region. Steve would fill me in on what he knew about Hamburg in general. We admired Hamburg's many fine old heritage buildings as we drove to the docks. We wound our way around the central lake with its magnificent fountains, and soon found ourselves at the harbour side. The docks area of this large German city is one of its busiest regions, and houses a great diversity of shops, cafes and trading businesses. As we walked along the waterfront, we discussed the location.

'Port cities are usually dens of iniquity when it comes to smuggling,' said Steve. 'Large boats coming and going all the time, plus the close proximity to other European countries, including Russia, or open sea routes to Asia and America, all make for many opportunities.'

A few blocks into our walk we came across a shop selling fossils so we decided to take a quick look inside. The dealer's name rang a bell, and I recognised it as one of those supplied to us by our German colleague. The shop had several rooms, each with glass cabinets filled with many fine specimens. (Not unlike most of the fossil and mineral shops I have visited around the world.) Large hollow rocks, filled with beautiful amethyst crystals, immediately caught my eye. Most of the specimens in the shop were the regular raw material of the trade—mineral specimens, trilobites from

Morocco, fossil wood from the United States, fossil fishes from Wyoming, even dinosaur bones from Montana. Then we started to notice the not so legal specimens. A stunning fossil bird from the Cretaceous of Liaoning, China, *Confuciusornis sanctus*, could be yours for a mere DM15 000, a beautiful *Mesosaurus*, a complete skeleton of a small marine reptile from Brazil (around DM4200), a range of dinosaur eggs from China, some meteorites from Australia. Steve pointed out the rare blue wood from Wyoming, all of which comes from one site on State-owned land. It is illegal to sell or trade this material, although it is permissible for people to collect small samples for private ownership. 'Smugglers regularly get in and rape the site,' he said in a low voice.

I asked the woman in the shop if there was a catalogue, but apparently one wasn't available. So I took the bold step of asking her if I could take some photographs of interesting items.

'But of course.'

I took out my camera and aimed it straight at the Chinese fossil bird. I shot off an exposure. Damn, the flash didn't work. Nervously I lined up a second shot. The moment the flash went off I heard someone enter the room and start speaking harshly to the saleswoman. It was the shop owner. Steve gave me the eye—it was time to depart, quickly.

I thanked the saleswoman and we left, listening to her being berated in the background. Steve's knowledge of German confirmed that the owner wasn't happy about us taking photos.

'Did you see the closed circuit television?' Steve asked.

'No,' I replied, feeling pretty damn stupid for not seeing the bleeding obvious.

Our van was waiting for us, with Alan at the wheel. We were ready for a quick departure, but were headed down a one-way lane clogged with traffic, so we had to slowly back up and head out past the shop again. The saleswoman was standing outside, smoking nervously and keeping a close eye on us. She didn't look very happy.

I realised at this point that, as good a film director as Alan might be, he's a lousy getaway driver.

Our task for the next day is to meet with a German dealer who has been implicated in illicit activities in the past, but never caught red-handed with the evidence. We are going to interview him undercover, posing as wealthy businessmen wanting something 'special' for one of our wealthy clients. We hope that during the interview we will learn something about the trade in dinosaur footprints or rare fossils from Australia, perhaps even hear something about the illegal side of the European fossil trade. Steve warned me that we have to be very, very careful. If someone can deal illegally with the Russians, get caught, and manage not to get arrested, then he must have either some powerful backers or connections in high places. Someone or something, probably money, had got him off the hook when he was arrested once before.

We dress in our finest suits, neatly pressed, with shirts newly ironed. Steve hands me one of his props—an expensive Rolex watch—and tells me to wear it loose on the wrist, flaunting it, making sure the German dealer sees it. Quality shoes are also a must.

'They always look at your watch and shoes to see if you really have money,' he says with a smile. I can see he's done this many times before.

Steve and I are both wired, with hidden microphones beneath our shirts, and the crew prepares a hidden briefcase camera. We have to fix on a plan, so we can control where the interview takes place. We don't want to have to leave our hotel, so the film crew can operate discreetly in the background. Steve comes up with a plan. I will be out when our man arrives, so that Steve can offer to have coffee with him in the hotel lobby. Then, on a signal from one of our crew, who will be sitting some way away from Steve and operating the briefcase camera, I will approach the hotel, see Steve in the lobby, and then make my entrance.

It's 2.30 pm, and our meeting is scheduled for 3 pm. I go across the road and sit in our hired Mercedes eight-seater van which is parked across the road from the hotel entrance, under some trees. Our cameraman is already stationed there with the large camera and zoom lens, ready to try to catch the dealer approaching the hotel. I wait with him and warn him each time I see someone approaching the hotel.

'There's someone coming.' He immediately starts filming the man walking along the path. But he's not our man, he walks right past the hotel. Then, instinctively, I know that *this* is our man, walking briskly up to the hotel. He is carrying some colour brochures under one arm. He is filmed approaching the hotel, and as soon as he reaches the entrance he turns into the lobby.

Once inside the man asks at the reception desk for Steve, and the receptionist directs him over to Steve's table. He walks over to Steve and introduces himself, all the while being filmed by Alan in the background, pretending to be a tired traveller—he has his pack, briefcase and papers spread out around him, and is sitting quietly on the sofa reading a guide book. Then he reaches for his mobile phone.

I wait impatiently in the van, fiddling with my mobile phone, which suddenly rings; the signal that it's time for me to enter the hotel. I walk straight across the busy road as there is a break in the traffic, and enter the hotel. Inside the lobby I glance around and then see Steve with the dealer. Steve gets up, shakes my hand, and introduces me to him.

He is a thin man, balding on top but with lank, longish hair. He has deep blue eyes and a moustache and smiles as he shakes my hand. Steve asks me how the business down town went.

'Oh, I took care of all that,' I say, smiling easily as I sit down.

On the table in front of us are glossy pamphlets from a US auctioneer. The dealer tells us about all the exquisite fossils that come up for sale on the international market. Enthusiastically he points out the whole complete ichthyosaurs from Holzmaden, large dinosaur bones from the USA, complete fossil mammoth skulls from Europe, whole cave bear skeletons from Russia. He claims that he can get us anything special we might want. When we press him, he turns the conversation around to what we want. He tells us that he will be our agent and can provide anything we desire . . . from Russia, South America, even China. We ask him about Australia, and he claims to have contacts there as well. As he talks, pointing out more and more wondrous (but highly expensive) items in the catalogue, I try to act dumb. He points out a superb fossil crinoid (sea lily) and I ask, naively, 'What is that—a fossil flower?'

He also points out a beautiful *Kueichosaurus*, a small marine reptile from China, most of which are about 20–30 cm in length.

'That looks good,' I say. 'Is it a big specimen?'

'No, only small, but exquisite,' he replies.

While Steve continues to chat to him I sit quietly and glance through the catalogues. I spot a specimen that would never have been granted an export permit from Australia— a rare fossilised starfish from the Gascoyne Junction region. It's called *Asterias* in the catalogue, but the starfish from this site have never been formally described. Professor Andy Gale from Greenwich University Geology Department, a world authority on fossil starfish, visited us in Perth some years ago and marvelled at the Permian starfish we had from this locality, all of which were certainly new species or genera, and therefore of unique heritage value to Australia. No undescribed fossil can leave the country, as it may be unique and never found again.

'Say, John,' Steve interrupts my reverie. 'How much did that colleague of ours, Wilson, spend on fossils for the foyer of his new office?'

'About $700 000,' I reply calmly.

'Yes, that's right,' says Steve, who then informs the dealer that Bill Gates had filled the entrance to his house with hundreds of thousands of dollars' worth of fossils.

Steve tells him that he wants him to be his 'main man' for advising him about fossil purchases. He becomes very excited at the prospect of doing future business with us.

Eventually, the hidden minicam runs out of film so Alan, seated on the sofa behind us, must make a quick exit before the camera starts to rewind. We are just about finished anyway, so make moves to end the meeting. We shake hands and tell him that we'll be in touch, before walking him out to the front of the hotel so our cameraman can film him with us. He then walks across the street, straight towards the van, right at the cameraman who is filming him. At the last minute he turns away and walks briskly down the street.

Later we discussed the information that came out of the interview, all of which was caught on tape. Steve explained that this kind of action would not be considered 'entrapment' in the USA, where entrapment implies that someone has had information forced out of them. In our case, the dealer talked freely and openly, and gave us whatever information he wanted. He just didn't know our true identities. Before I arrived, he had been explaining to Steve how he had in the past procured fossils from Russia, telling him that most academic staff in that country are so in need of money that it's quite easy to get specimens out of the universities or museums. He also explained, in my hearing, how he could get any fossils we wanted from Argentina, explaining that he was about to take over a major part of the market trade in South American fossils. I think he said that he was 'making moves to control fossil sales and exports out of that country', but I could be mistaken.

That evening we relaxed in downtown Hamburg in a small cafe, and partook of a local beer or two. It was an early night for us all, though, as we had managed little sleep the night before on the plane from Australia. I think I had survived the day on a combination of adrenaline and coffee.

'How did you feel about today, John?' Steve asked me later.

'It was OK, but a bit scary at first, realising that one slip-up could make him suspicious of us. But it got better as I relaxed and went with the flow.'

'You did just fine, John,' he replied, flashing a broad, toothy smile.

We departed Hamburg the next day for Frankfurt, where we had an appointment to visit one of Europe's oldest and largest natural history museums, the Senckenberg Museum.

In Frankfurt

7

After long deliberations, and at least one unsuccessful purchase effort, the Trustees finally approved the recommendation of Sir Richard Owen and authorised payment of £450 from the 1862 budget of the Museum and £250 from the following year's budget to purchase the *Archaeopteryx* specimen and a number of other Solnhofen fossils in Häberlein's collection. For the total sum of £700, the British Museum acquired an outstanding collection of 1,703 specimens, but the prize, of course, was the feathered fossil that is commonly referred to today as the London specimen of *Archaeopteryx* (John H. Ostrom, *Discovery*, vol. 11, no. 1, May 1975).

A sketch of the famous *Archaeopteryx* specimen sold to the British Museum for £700. The specimen is just under 50 cm in length.

The pattern we see emerging in Europe during the 1820s–1870s is of fossil collectors amassing large and significant collections then selling them for high prices to leading museums. Museums thus played a vital role, by sponsoring the private collection and trade of fossils, and also by acquiring and protecting the most important specimens. The system still tries to work along those lines, but it fails when prices become far too high for modern museum accountants to consider.

Germany is the home of one of the most famous fossil sites in the world, the Solnhofen limestone quarries, where *Archaeopteryx*, the earliest fossil bird, was found. The first almost complete skeleton (albeit without the skull) of *Archaeopteryx* was found by palaeontologist Karl Frickhinger's great-grandfather Albert, who saw the slab containing the bird propped up against the wall of an inn. He eventually passed on the specimen to Carl Häberlein in 1861 (Frickhinger 1994). Häberlein was a physician who collected fossils, often accepting them in lieu of payment of his bill. He treated many of the local quarrymen who worked in the flat lithographic shales around Solnhofen. Over time Häberlein amassed a huge collection of fossils, but when his daughter wanted to get married he needed to raise a dowry, so the fossils went up for sale. Among them was the rare fossil bird with feathers.

This was one of the most significant fossil sales in Europe. The £700 purchase price had to be split over two years of the British Museum's budget. In the mid-nineteenth century the average working-class person in England would have earned £15–40 per annum, so £700 was a considerable fortune.

Tuesday 11 September. The Senckenberg Museum has some of the most amazing fossil specimens I have ever seen. In the entrance foyer is a complete skeleton of a giant marine reptile, the ichthyosaur *Eurhinosaurus*, some 4–5 metres long, lying on its side in a single slab of Holzmaden shale. There is also a huge crocodile and a magnificent slab of black shale showing crinoids (sea lilies) attached to floating debris by their stalks.

The hall of dinosaurs dominates the central space. All the main groups of dinosaurs are represented: *Tyrannosaurus*, *Iguanodon*, *Triceratops*, *Stegosaurus*, *Euoplocephalus*, *Diplodocus* and *Supersaurus*. Hanging from the roof is a mounted skeleton of the pterosaur *Anhanguera*. Below this main chamber is a smaller room holding a superb collection of real fossils, and there, in a large glass case, is one of the world's only known mummified dinosaurs—an *Edmontosaurus* with skin impressions preserved. In fact the skin appears to be preserved around each of its hands. It is bent over in a post-mortem crucifixion position, its arms outstretched.

We started with a general tour of the museum's galleries, including their world-famous Messel gallery, featuring incredibly well-preserved reptiles, birds and mammals from the Messel oil shales. Then we took a behind-the-scenes look at their preparation lab, where we were shown remarkable specimens, including clutches of dinosaur eggs from China, large crinoids from Holzmaden in Germany, a spectacular fossil whale jaw from Egypt and a tortoise from Liaoning, China. After this tour we interviewed Professor Dieter Peters about his research.

Fossil birds from China

Over the last few years the Senckenberg Museum has been featured in various scientific journals as one of the few museums openly acquiring Chinese heritage fossil material. The first case in question was the purchase of several specimens of *Confuciusornis*, a primitive fossil bird from the Early Cretaceous of China. The species was formally described in 1995, and only a few specimens have come onto the open market. The first specimen turned up in 1993 in the United

States, when a Chinese businessman tried to have it auctioned by Phillips in New York, two years before the bird had been described and formally named. Even so, the bird was recognised as a valuable heritage item, and Phillips' consultants told the auctioneers not to touch the specimen, so it was refused. In December 1996, however, one year after it had been described, another specimen would be auctioned, despite the fact that the restrictions of Chinese heritage laws meant that it had to have been exported illegally. The Senckenberg Museum has purchased a number of Chinese fossils birds and other fossils, and has been researching these specimens with Chinese scientists as co-authors.

Professor Dieter Peters is a well-known palaeornithologist, a specialist who studies bird evolution. He has worked with Dr Ji Qiang from the National Geological Museum of China on the *Confuciusornis* specimens purchased by the Senckenberg Museum. The new specimens were clearly of great scientific interest and generated important new research observations. Their first paper, published in 1998, described the skull in detail, showing it to have a typical diapsid (having two temporal openings) skull (Peters & Qiang 1998). This paper also demonstrated that *Confuciusornis*, although birdlike in many features of its skeleton, had a dinosaurian type of skull. The specimens they described were held in the collections of the Natural History Museum of Vienna (number z0112) and the Senckenberg Museum (number Av 416). A second paper entitled 'Had *Confuciusornis* been a climber?' illustrates two additional specimens (Av 412 and 423) registered in the Senckenberg collections (Peters & Qiang 1999).

Professor Peters was most helpful, offering to show us their best *Confuciusornis* specimen, purchased from a local fossil dealer just after the first specimens of the bird were formally described (Hou *et al.* 1995). He disappeared into the storage area and returned a few minutes later, with a magnificent specimen of *Confuciusornis sanctus*. According to Professor Peters, the Senckenberg Museum's new specimens apparently came with papers declaring that the specimens were legally exported, although the paperwork was in Chinese and the museum does not have an exact transcript. When asked about the legitimacy of the specimens, Peters replied:

'I saw the documents. I can't read Chinese but I was told that there were official permissions, from authorities, from Chinese authorities, given for this particular specimen.'

He also told me that the specimens had been purchased from a reputable German dealer.

Our discussion ranged over other issues about fossil smuggling and ownership. Although most, if not all, important vertebrate fossils from China are known to be illegally exported, the main issue that Chinese scientists are, rightly, worried about is where such specimens would end up if bought by private individuals, rather than a major natural history museum. Following this argument to its logical conclusion, it is easy for international museums to justify purchasing such scientifically important specimens without proper provenance, to make sure they are properly conserved. Furthermore, specimens in the Senckenberg Museum are available to any scientist wishing to study them.

The new fossil birds from China are extremely significant in the big picture of bird evolution. *Archaeopteryx*, from the late Jurassic of Germany (around 150 million years old) has a reptilian type of skeleton, but with extended arms and fingers with feathers. (The Senckenberg Museum has an excellent display of all the major specimens of *Archaeopteryx*, mostly replicas but one or two original specimens.) *Archaeopteryx* has teeth and lacks a pygostyle, the shortened tail bone that characterises all modern birds. Some scientists regard it as the first bird, others consider it to be a feathered dinosaur. *Confuciusornis*, *Sinornis* and *Liaoningornis* from the Early Cretaceous of

Liaoning Province, China (120–130 million years old), show various steps in the evolution of more modern birds. *Confuciusornis*, with its dinosaurian skull, has a beak without teeth and a pygostyle. *Liaoningornis* has a well-developed keel on its sternum and an expanded rib cage, suggesting that it had the advanced respiratory system needed for sustained flight. *Sinornis* is regarded as one of the earliest true birds. It has a well-developed sternum and a complex wrist joint that assisted flight strokes of the wing. The Chinese sites have yielded two other primitive Cretaceous birds, *Chaoyangia*, possibly an early toothed wading bird, and *Cathayornis*, a small perching bird quite similar in its skeletal anatomy to *Sinornis*.

My point is that with all these varieties of fossil birds coming out of the Liaoning sites (and equivalent deposits), the recent flood of specimens on the international fossil market could easily see new species emerging, but not being recognised. Specimens of new genera could quite easily be sold to private collections and never be studied. We will never know if this has already happened, nor will the holotype specimens ever find their way back to museums in China, where they belong.

We had finished our interviews at the Senckenberg Museum and were having coffee with Professor Reischel of Karlsruhe in the administrative offices when Dr Peters telephoned with the shocking news of the terrorist attacks on the World Trade Center in New York and the Pentagon in Washington. Chaos has broken loose in the free world, and it will now be very difficult for us to fly to the USA in the next few days, as we had planned.

The German police and fossil seizures

Wednesday 12 September. We visit Chief Superintendent Eduard Amrien, a member of the environmental division for the state region. His jurisdiction spreads over the whole country on matters of fauna, flora, fossils or other environmental issues.

Steve chatted with Amrien in German, explaining that he too was a law enforcement officer, a specialist in fossil-related crime. Amrien seemed very surprised that fossil-related crime in the USA was serious enough to warrant its own specialist policeman.

Amrien explained German legislation pertaining to fossils. As far as he knew there were no laws protecting fossils per se, only laws related to the land the fossils were found on. The landowner also owned any fossil sites on his property and could do as he liked with the fossils they contained, except in special cases such as Messel, where a site is designated by UNESCO as a World Heritage site. In such cases stricter measures are put in place, to protect the site from public access. In short, fossil collectors in Germany can collect anywhere they like as long as they obtain the landowner's permission. In some cases popular fossil sites can be entered after a small entrance fee is paid.

Amrien outlined his role in the 1993 seizure and return of fossils illegally exported from Australia to Germany. Acting on reports that suspect Australian fossils were on display at a trade show in Japan and had been purchased by a German dealer, and in response to an official diplomatic request from the Australian government, the German government issued a search warrant, enabling the German police to raid the private dwellings of a Frankfurt fossil dealer, Joachim Karl, and his preparator, M. Walter, a geology student. The police confiscated a number of fossils taken from sites in Australia: perfectly preserved crinoids from Western Australia, fish in nodules from Gogo in Western Australia, large trilobites from Kangaroo Island in South Australia and rare Ediacaran fossils from the Flinders Ranges of South Australia.

The two Germans had not broken any laws, so the fossils were confiscated but no charges laid. The specimens were then brought back to Australia as evidence in the resulting trial in 1997 (see Chapter 4).

We thanked Chief Superintendent Amrien and left police headquarters. We decided to take the opportunity of further exploring Frankfurt's fossil shops to find out what kinds of fossil specimens could be bought over the counter in Germany. Our first visit was to the fossil and mineral shop of Joachim Karl. The small shopfront bristled with neatly presented small specimens. The shop displayed no material that could be identified as being illegally imported, only a few fossils from regular sources like Morocco and the commoner species of Green River fishes from Wyoming.

Later that day I visited another fossil shop. This shop was packed with many amazing fossils, of high quality and rarity. Unfortunately, I only had a few minutes to look around the shop as I had to meet up with the others. Although we were scheduled to depart Germany the next day, I dearly wanted a more detailed look at the shop if time permitted.

Thursday 13 September. We are scheduled to be on the early flight to Denver. But on reaching the airport, we are told that Denver Airport is closed, so we are now on a waiting list for a flight out either on Friday or Saturday. We book into a hotel in downtown Frankfurt and then walk downtown to get some lunch.

We ate at one of the colourful street cafes in a tree-lined mall. While munching away on Bratwurst and fries, washed down by strong German coffee, Steve noticed a woman go by. As we were finishing our meal, he saw her again, and then told Alan and myself that she had actually walked by us three times since we had sat down. Was she tailing us? As we should have flown out of Frankfurt that morning, it must have been a coincidence that she had picked us up in town. But, Steve mused, we could have been tailed throughout our stay in Frankfurt, so we wrote her off as coincidence.

Steve and I then went back to the shop I had visited yesterday. Steve was wired with a hidden microphone, and I was carrying a briefcase camera to try to capture the full range of specimens on display. I reminded Steve that in Germany it was not illegal for this material to be on sale, so the dealer wasn't breaking any laws. We were just noting what material was being taken out of other countries, presumably illegally.

I was immediately taken aback by the large number of beautifully-prepared, enormous fossil fishes that adorned the walls—a spectacular array of Green River fishes from Wyoming, local specimens from Solnhofen and rare *Acanthodes* specimens from Odernheim, as well as superb large fishes from the Santana Formation of Brazil. Some of the most notable specimens were two stunning *Heliobatis* rays from Wyoming, a *Rhinobatis* ray from Solnhofen and a very large predatory fish, about a metre long, in an orange nodule from Brazil. These specimens were priced in the thousands of marks range.

Steve asked the proprietor about the Green River fishes, and the man looked at him suspiciously.

'Haven't you ever seen these fishes before?' (They are sold in many shops around the world.)

Steve replied that yes, he had seen them, but these were of exceptionally good quality.

The proprietor then explained that he has been on excavations in the USA, and that he had obtained some of his specimens from illegal land in Wyoming. He said, smiling, that he had been warned that the police could arrest people 'over there' (the USA) for digging on that land, emphasising the dangers of his trip. This was probably part of the rationale behind his exorbitant prices.

Steve pointed out one slab containing many small fish, *Gosiutichthys*, a small herring. According to Steve, these all originated from one site on State land, so they had to have been taken illegally.

A woman wearing a black leather jacket then entered the shop. Steve told me to get her picture, as she was the same woman who had walked by our lunch table three times. Steve never missed these 'coincidences'. By this stage I was getting a little nervous. Steve continued to chat while I looked around at the jewellery displays, trying to ignore the fossils. I switched the briefcase to my other hand so that it was pointing backwards at Steve and the proprietor. Steve asked him about his amber specimens with insects inside. He told Steve that he went to Russia himself to get the material.

I decided to deflect attention from myself by buying something for my daughter, and picked out some pretty coral earrings, from the same case as the amber. The shop assistant, who had seen me in the shop yesterday, took the earrings out of the glass case and wrapped them up for me. I took the package and popped it in my pocket. I thought it was high time to leave, as we had been in the shop for far too long. Steve whispered to me that he had identified at least four security cameras, monitoring our every move. It was possible that microphone sensors were fitted in different parts of the shop. It was definitely time to leave. Back in the safety of our van we discussed the episode, which had been an eye opener for both of us. Steve was amazed that the proprietor would brag to a total stranger about stealing fossils from government land. He might be safe from prosecution in his own country, but the chances were that he would return to the USA one day. When he did, he would be watched closely.

Steve smiled at us, but his eyes were serious as he said, 'He had better take care if he ever visits Wyoming again.'

A famous German fossil site

Saturday 15 September. Steve calls Broome Prison again, but Michael Latham still isn't available. We visit the famous Messel fossil quarry near Darmstadt.

One of the world's most important sites, the 49-million-year-old oil shales of Messel, preserves a great diversity of insects, plants, fishes, frogs, reptiles, mammals and birds. The preservation of these fossils is truly extraordinary and like no other site on Earth. The mammals often have their fur preserved, stomach contents intact, even muscle tissue preserved as microbial replacements. Some of the fossil insects even have colours preserved on their delicate carapaces.

Until the 1970s the Messel quarry was in active use as an oil shale resource, but now it has been designated as a World Heritage site by UNESCO. Various academic groups slowly work the site, with permits. We were permitted to watch as students from one of the universities quarried slabs of black shale and then split them with chisels and small knives to find the fossils they contained. I asked them about security. The entire quarry is heavily protected by wire fences topped with barbed wire, and large gates that are locked when the quarry is not being worked. We were also assured that security guards patrol the site on weekends to keep out unwanted visitors. It appears that Messel is indeed a well-protected site that is used for scientific excavations only. Few Messel fossils are available on the international fossil market. Most of the rare and scientifically valuable Messel material is housed in various natural history museums and university collections in Germany, with a fair representation of some specimens in large overseas museums.

We had lunch in Darmstadt, famous as the home of Frankenstein's castle, after which we rang Digby Macintosh in Broome to see how he was, and to find out if anyone had questioned him about our visit. He seemed OK on the phone, and told us that he hadn't heard any new information about the footprints.

Sunday 16 September. I walk around the Senckenberg Museum to look at the exhibits at my leisure.

I spot a large slab of *Jimbacrinus* crinoids from Western Australia (not labelled as coming from Western Australia, just as 'crinoids'). It's a beautifully prepared slab, quite professionally done. There is also a large slab of shale from Liaoning, China with a beautiful fossil tortoise. I wonder if this is possibly an undescribed new species, as no identification is given on the label. Several Green River stingrays from Fossil Butte, Wyoming, are well displayed.

I was very impressed by the displays at the Senckenberg, truly one of the world's greatest natural history museums, although I haven't come to terms with the ethics of their methods of obtaining specimens. As a museum curator myself, I tend to agree with their attitude, that fossil specimens of great rarity and scientific importance which come on the international market should be purchased by museums that will conserve them and make them publicly accessible for research.

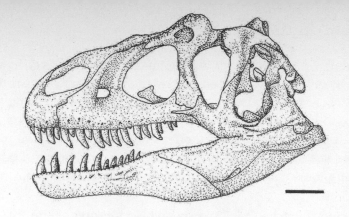

Denver, Utah and South Dakota

8

Over the next ten years, the Wyoming fossil rush was as exciting if not quite as lucrative as the California goldrush had been over 30 years earlier. Marsh and Cope continued to mine Como Bluff and other fossil fields in the region for dinosaurs. They apparently thought nothing of spying, bribing, intimidating, and stealing each other's employees. When Marsh's man Reed finished digging at a site, he destroyed all remaining bones so that Carlin could not come in and ship them to Cope (Fiffer 2001).

Sketch of the skull of *Allosaurus*, a well-known Jurassic dinosaur from Utah. (Bar scale is 10 cm)

Monday 17 September. We are finally able to travel to the USA, after being wait-listed on the flight and hanging around the airport for five hours. Our original aim was to attend the Colorado Mineral and Fossil Exposition in Denver to check out the latest range of fossils offered for sale. We have now missed the show, so will have to content ourselves with interviewing local fossil dealers, if we can get their permission.

We ring Charlie McGovern, who operates 'The Stone Company'. He sells fossils of all kinds and also acts as a consultant for fossil valuations, setting up museum exhibitions and so on. He agrees that we can interview him tomorrow at his house, where he stores most of his fossil collection.

With his long, greying hair and grizzled beard, Charlie McGovern looks a bit like the Gandalf of the fossil traders. Like the mythical wizard, he has been in his trade for a number of years and clearly loves the business. He has been featured in *National Geographic* magazine, and has supplied fossils to several of the world's major museums. Charlie and I first crossed paths a few years ago when he acted as the main fossil-procuring agent for a new natural history museum in Gamagori, Japan. I was involved in the same project, checking their scientific information on fish evolution.

The first thing I noticed on arriving at Charlie's home was the fine examples of polished petrified wood slabs which adorned his walls. His office is a treasure trove, with some specimens just lying around, other special items displayed on stands and still rarer objects inside a glass case. Alan and I interviewed Charlie about the likelihood of the Australian fossils coming up for sale in the USA. We also questioned him about dinosaur footprints and whether or not there is much of a market for them.

His answers were direct and to the point. He had no knowledge of ever seeing the Australian prints at US fossil trade fairs, nor did he think they had come to the USA. He told us that dinosaur footprints from the Connecticut Valley regularly come up for sale at fossil shows, but generally sell for small amounts, say US$100–$500. There isn't much of a market for dinosaur footprints.

'Near is dear,' he advised us.

'What do you mean?'

'Well, they [the stolen prints] are more likely to be valuable to someone in Australia who knew about them and specifically wanted them. Specimens like that generally don't travel far. My guess is that they are still somewhere in Australia, most likely that they went to a local collector from the same area.'

Charlie then asked me to sign his copy of my book *Rise of Fishes*, before asking me if I knew Bob Bakker, the famous dinosaur expert. I had met Bob a few times at palaeontological conferences, so Charlie rang him up and asked him to come round. Bob is one of palaeontology's most outlandish characters. His long hair is topped with an ever-present cowboy hat, and he dresses like someone who has just moseyed in from a ranch, rather than a fossil laboratory. Despite his rugged appearance, Bob is one of the greatest living palaeontologists. He led the renaissance of dinosaur palaeobiology back in the early 1970s, and his books have changed the way we look at these animals.

When Bob arrived we shook hands, and joked about the film series.

'What do I think about the fossil trade? They are all a bunch of brigands!' he said, smiling.

We then interviewed Bob in front of the camera. He told us that fossils' most dangerous enemy is nature—erosion. Good specimens are being lost all the time because there simply aren't enough people who can find them and dig them out. Well-trained fossil dealers who can excavate specimens are thus responsible for saving many scientifically important specimens which would otherwise be lost to science. He used the example of Sue, the *Tyrannosaurus rex* excavated by the Black Hills Institute headed by Peter and Neil Larson, to illustrate his point.

Sue's story

Sue, an almost perfect *Tyrannosaurus rex* skeleton, was found by fossil hunter Sue Hendrickson on the morning of 12 August 1990, on a ranch in South Dakota, and later excavated professionally by Peter Larson and his team from the Black Hills Institute. It took seventeen days of careful excavation to get the skeleton out. Larson then told landowner Maurice Williams that they had found a really good specimen on his land and would pay him US$5000 for the specimen. To further complicate matters, the Cheyenne River Sioux Indians claimed on 29 August in the *Rapid City Journal* that the dinosaur had been taken without permission from reservation land, an act in direct contravention of Federal law. On 10 November Larson received a letter from Williams stating that he hadn't sold the specimen to the Black Hills Institute, just given them permission to remove it for cleaning and preparation for sale. The *Rapid City Journal* also reported that Williams had entered into a partnership with the Sioux to determine the dinosaur's future.

At 7.30 am on 14 May 1992, 35 law enforcement agents, including the FBI and local sheriffs, descended upon the Black Hills Institute. Larson was handed a writ alleging crimes such as the felony of stealing from government and tribal lands, as well as violations of the *Antiquities Act* of 1906. The town rallied round to save Sue from being confiscated. Larson had arranged for the dinosaur's skull to go to NASA for a CAT scan to reveal its three-dimensional internal anatomy. US Attorney Schieffer would not allow the specimen to go, however, and put motions in train to have the whole skeleton taken away. It took the government agents only three days to pack up the skeleton, despite the countless thousands of hours of work it took to prepare it. Several well-known palaeontologists, including John Horner, Phil Currie and Bob Bakker, went public with their outrage at the government's rough treatment of this unique specimen.

Sue's story was long and involved with various court cases and appeals, and is well documented in Steve Fiffer's excellent book *Tyrannosaurus Sue* (Fiffer 2001). Finally, on 15 December 1993, Judge Magill used the South Dakota property law to hand down his decision on the ownership of Sue. He concluded that the fossil was 'land', and that for millions of years the bones had been buried as a mere 'ingredient' of the soil that the United States held in trust for Williams.

Criminal charges were laid against the Larsons and Peter Larson was eventually sentenced, quite harshly in view of the charges, to two years for retaining (buying) fossils valued at less than US$100 taken by a third party from Gallatin National Park, plus two counts of customs violations with respect to taking money out of the country. He had been convicted of only two counts out of 33 felony charges. The US government had spent millions of dollars on the case, but the Larsons were acquitted of the main charge, that of being involved in a conspiracy to steal fossils from public lands. The jury had not seen them, or their institute, as being in any way fraudulent. (I first met Peter Larson and his brother Neil at the inaugural Dinofest Conference in Indianapolis in 1994, and was impressed by the level of scholarship Peter exhibited in his presentation on the anatomy of the new *Tyrannosaurus* skeleton. Many of the other palaeontologists also commented upon his good work. He had identified wounds in Sue's skeleton and had made detailed measurements to determine sexual dimorphism in *Tyrannosaurus*, arguing that Sue was clearly a female dinosaur (Larson 1994).)

On 4 October 1997, Sue the *Tyrannosaurus* went under the auctioneer's hammer at Sotheby's in New York. It was sold to the Field Museum of Chicago, one of the world's foremost natural history institutes, for the princely sum of US$7 600 000. With Sotheby's commission, the real price paid for the dinosaur was US$8.36 million (in Australian dollars in 2002, that's around $16 million). The funding to purchase Sue for the Field Museum was put up by McDonalds and Disney. The second-highest bidder was the North Carolina State Museum, which offered US$7.5 million for her, while the third-highest bidder was a private foundation which intended to donate the specimen to a natural history museum in Florida if they were successful. It is cheering to note, therefore, that Sue would have ended up in a legitimate US museum regardless of which bid won.

Sue is now on display in the Field Museum of Chicago, for all to marvel at.

Wednesday 10 September. At Bob Bakker's invitation, we spend the day at a famous dinosaur fossil site, Como Bluff east, to get some perspective on the rarity of good dinosaur bones, and on how delicate and time-consuming is the task of excavating them.

The three-hour drive from Boulder took us through some very scenic country. We followed Bob in his field vehicle, a small, rusty and aged Datsun. The back seats and passenger's seat were filled with an assortment of digging tools, bags of plaster, rolls of hessian, bottles of superglue and other dinosaur hunter's paraphernalia. Along the way I tried to figure out the local geology, my eyes lighting upon rugged escarpments that looked ideal for fossil hunting. Two hours later we pulled into the town of Rock River, Wyoming. We all walked into a local cafe, Longhorn Lodge, for lunch.

'Best chilli burgers in Wyoming,' Bob said as we walked inside. A couple of crusty oldtimers sat talking and drinking coffee. As we ordered our meals in broad Aussie accents, they stared at us as if we were from another planet.

After lunch we followed Bob along a dirt trail that wound through gently rounded hills of grey-, white- and buff-coloured sediments. The late Jurassic Morrison Formation, about 145 million years old, is arguably one of the world's richest dinosaur-bearing rock units. The site Bob was currently working was a small quarry, formed by levelling out the cliff halfway up a low hill of light grey sandstone and shales. A tarp was stretched across a flat excavated area that had large white blobs of plaster scattered across its floor.

'They're all dinosaur bones,' Bob told us. Each one had been carefully excavated and plaster jacketed, and some were now ready to go back to the lab for further preparation. Here before us were the scattered remains of the great Jurassic giants—the long-necked *Apatosaurus* (or *Brontosaurus*, as Bob prefers), *Camarosaurus*, the carnivore *Allosaurus* and the bone-plated *Stegosaurus*. Working here over the last twenty years or so, Bob's field teams had discovered 50–60 different fossil sites. Many hundreds of dinosaurs' bones had been carefully excavated, with painstaking drawings made of their positions, and measurements of their three-dimensional orientation in the ground. One of Bob's doctorate students had been working on the diagenesis of the site, studying how the fossils had formed within the sedimentary layers, and how the decomposition of the dinosaurs' flesh had chemically altered the sediments around them. She told us that the dinosaurs' decomposition caused 'saproglabritic' nodules to form in the soft mud, through gas bubbles escaping from the flesh. The gases inherent in the sediment sometimes caused bones to be uprighted within the soft, muddy layers, before they hardened to sedimentary rock.

Each site at Como Bluff east is different, and each tells us something new about the world in which dinosaurs lived.

One locality was interpreted as a feeding site, where mother theropods dragged lumps of meat back to their young. The main signpost for this was the presence on the site of many shed teeth, some only 4–5 mm in length, obviously from new hatchlings. Other sites are dominated by aquatic fauna such as turtles and fish. Every site in the whole region is of great scientific importance because when collated, the information they contain provides us with an overall picture of a complex terrestrial environment in the late Jurassic. This area of North America was then an isolated island, with a number of endemic fauna and flora, much like Australia today.

As I lay on my side in the dirt, picking away for hours at the sediment surrounding a large *Stegosaurus* vertebra, I began to realise how much was involved in collecting a single dinosaur bone. To record its position in the sedimentary layer properly, excavate it without damaging it, plaster jacket it, transport it back to the lab and then prepare it was clearly a time-consuming job, requiring considerable skill. In many cases the bones start to dry up as soon as they are exposed to sunlight, causing the surface layers to crack and flake off. We had to pour superglue liberally over the surfaces of the newly exposed bones and let it soak right in, to combat this. Some cases of fossil poaching I had heard of from Steve involved the offenders literally ripping a dinosaur skeleton out of the ground in a day or two. That sort of rough treatment would not be good for the bones.

I don't know where the afternoon went, while I was lying next to the bones, listening to Bob chatter away about all things wonderful and Jurassic, but it was a very pleasant day. Among Bob's many tales was a memorable one about the first case of professional dinosaur rustling, a story all the more thrilling because it took place in the same region where we were digging.

The bone wars

No book about dinosaurs and fossil trading would be complete without some mention of the famous American palaeontologists Cope and Marsh. Most of us remember them for their infamous bone wars, each one racing to find and describe more species of dinosaurs (or mammals, fish, reptiles, whatever they could find) than the other. Charles Othniel Marsh, Professor of Palaeontology at Yale University since 1865, had men digging in the Como Bluff area since the first discoveries there in 1877, and his team had excavated many fine skeletons which were shipped by train back to New Haven for analysis and description. His arch rival, Edward Drinker Cope, was then a man of means living in New Jersey and based at the Philadelphia Academy of Sciences. He was a prolific research scientist, publishing enormous amounts of work (in 1872 alone Cope published some 56 scientific papers). In 1868 Marsh and Cope spent a week together hunting around in the Cretaceous rocks of New Jersey. Only one year later they were bitter enemies, but Colbert (1968), unlike other authors, doesn't attribute their rivalry to one particular event, but instead puts it down to their ambitious, competitive natures, and the fact that neither of them was over-scrupulous if the other stood in his way.

In 1879 Marsh's team had a particularly good find, the first complete skeleton of a theropod, a meat-eating dinosaur. Up to this time various bones, including skulls, had been found, but no complete articulated skeletons were known. Not a single museum anywhere in the world had a meat-eater in a fully mounted display. Under cover of darkness, Cope's men slipped into Marsh's pegged site and excavated the complete skeleton. It was shipped back to New York, but because the boxes were not properly labelled, and many of the specimens were just partial bones or debris material, the boxes went unnoticed for many years. In 1906, a few years after both Cope and Marsh had passed away, the boxes were opened and the magnificent skeleton of *Allosaurus* prepared and mounted for public display. Its reign as the world's largest meat-eater would be short-lived, however, as in 1908 the first complete *Tyrannosaurus rex* skeleton was found in Montana by Barnum Brown; it was erected at the American Museum of Natural History a few years later.

Cope sold most of his lifelong collection, comprising some 10 000 specimens, in 1895 to the American Museum of Natural History, for nearly US$32 000. The rest of his collection, sold later, fetched another US$29 000 (Boyce 1994).

Thursday 20 September. We move on from Boulder to Salt Lake City. After stringent security checks at the airport, we finally board the plane, which has numerous empty seats.

Each of our party has an entire row of seats, so I spread out and enjoy the flight, musing over the geology of the Rocky Mountains. Below me I can see a vast area of eroding sedimentary rocks, and I know that there are perhaps millions of fossils in these isolated regions that are weathering out, that need to be collected and studied if new pieces are to be added to the big jigsaw of life. Such sights encourage the spirit of fossil collecting, whether it is by amateur or professional. As long as someone gets the specimens before they weather away to dust.

Friday 21 September. We are in Salt Lake City, northern Utah, and today we will visit a private fossil museum. The North American Museum of Prehistoric Life, at Thanksgiving Point in Salt Lake City, opened to the public only a few months before our visit. It's a private museum, a tourist venture sponsored by businessmen who saw the commercial potential of dinosaurs.

Unlike the major State-operated natural history museums, such as the Utah Museum of Natural History (situated on the campus of the University of Utah), private museums have no ethical obligation to let foreign scientists study their specimens, or to loan specimens to academic institutions. Such museums can, at any time, dispose of their specimens as they wish.

I had a good look through the museum and must say that I was very impressed by its displays. The quality of fossil specimens, the accurate information supplied on the labels, the scenic backdrops and atmospheric lighting all showed that it was professionally put together, with an impressive budget for design and exhibition planning. One of my good friends in Perth, Travis Tischler, a renowned sculptor of prehistoric animals, was commissioned to make several models for the museum, and I was the palaeontologist who signed off on their authenticity. So, I reasoned, most of the scientific displays there would also have been checked out by other palaeontologists.

There were some very interesting specimens on display—Chinese dinosaur eggs, dinosaur bones from China and Mongolia, fossilised birds, such as *Confuciusornis* from Liaoning, a well-preserved original skull of a pterosaur (a flying reptile), from the Santana Formation of Brazil and skulls of mammal-like reptiles from South Africa and Russia. All of these might well have been bought legally in the USA, but they would have been smuggled out of their country of origin. There were also some spectacular dinosaur remains—original, complete skulls and partial skeletons of magnificent late Jurassic species which were labelled as new genera, despite the fact that the names had not been formally ratified through publication in acceptable journals (as designated by the International Code of Zoological Nomenclature). Rather than add to the existing confusion of taxonomic literature, I won't go into details here.

The museum is an impressive one, obviously enjoyed by members of the public. The specimens are fantastic, both originals and very well-made replicas, and Travis' excellent reconstructed models of prehistoric animals, such as the dinosaurs *Coelurus* and *Othneilia*. However, the museum can also be viewed as an elaborate extension of a very large fossil shop in which the best specimens are put into professionally-made commercial displays. There is nothing wrong with this concept—where the problem lies is that none of the specimens is in a publicly recognised registered collection. In other words, no holotypes of new species can be deposited there, nor can specimens be loaned to or accessed by other scientists for study, unless, of course, such requests are approved by the Trustees or Board of Directors. To my knowledge there are no professional palaeontologists to curate or conserve the museum's specimens. As a commercial venture, the museum's scientifically important specimens are, to my mind, only as safe as its profit margins. If the museum ever became unprofitable, the specimens on display could be auctioned to pay off its debts.

The many mounted skeletons of dinosaurs in the galleries reminded me of a current case in Utah, one of a poached *Allosaurus*.

A Utah dinosaur sold to Japan

In September 2001, newspapers in Utah publicised an upcoming lawsuit which alleged that the defendants stole the skeleton of a meat-eating dinosaur, *Allosaurus*, from a site on Bureau of Land Management property in Utah, and later sold the specimen to a Japanese museum for US$400 000. *Allosaurus* was a 12-metre long theropod, whose remains have been found at several late Jurassic sites throughout Utah and Wyoming. Although some species are well known, there are still ongoing scientific studies of *Allosaurus* which makes any new discovery an important one for scientists trying to figure out just how many different species of *Allosaurus* roamed around North America 180 million years ago.

The defendants, Barry and April James, claimed that they legally bought the skeleton from a Utah fossil dealer, and had the paperwork to prove it. Mr James, a teacher with a master's degree in vertebrate palaeontology, and his wife run a company called 'Prehistoric Journeys', which sells fossils and fossil replicas to museums. In addition to the charges of theft, Federal prosecutors filed a US$2.1 million civil suit against the Jameses, arguing that the couple hired someone to excavate the dinosaur skeleton. They contend that Barry James first heard about the dinosaur in 1991, when a Rocky Barney took him to the excavation site. There, the two discussed removing the fossil, including the legality of removing fossils from Federal land.

'A professional excavation by legitimate palaeontologists would have taken six months. Instead, amateurs using picks, shovels and wheelbarrows dug it up in nine days,' said Don Johnson, head of the FBI office in Salt Lake City. Investigators say the fossil is worth US$700 000, but that James bought it for a mere US$90 000 and sold it for US$400 000.

The Japanese museum that bought the specimen did not know it was an illegal export, and was not charged.

Other *Allosaurus* bones had previously been stolen from a very famous fossil site, the Cleveland-Lloyd Quarry in Utah. This quarry, first worked in 1927, was designated a natural landmark in 1967 and is now under the jurisdiction of the Bureau of Land Management. In September 1996, someone broke into the visitor centre at the quarry and stole specimens of *Allosaurus* as well as the large sauropod *Apatosaurus*. In October 1996, the Emery County Sheriff's Office and the Bureau of Land Management offered a US$5000 reward for information leading to the arrest and conviction of any person or persons involved in the theft.

Our visit to Utah was the perfect opportunity for me to interview Dr Scott Sampson. Scott is one of North America's up-and-coming young dinosaur palaeontologists. A veteran of many field expeditions, he cut his professional teeth excavating and studying the great horned dinosaurs in Montana, naming several new genera including *Einiosaurus* and *Achelousaurus*. His expeditions to collect and study Gondwana dinosaurs in South Africa and Madagascar have resulted in some spectacular finds which have proved crucial to our understanding of dinosaur biogeography. Among his teams' discoveries was one of the most perfectly preserved theropod skulls found to date, *Majungatholus*, which featured on the cover of *Science* magazine on 15 May 1998 (Sampson *et al.* 1998).

Scott had only just taken up the position of Vertebrate Palaeontology Curator at the Utah Museum of Natural History. He gave me a tour of his collections, pointing out all the great material he had to work with that had been collected from Utah. He then showed me some of the original bones collected from the new sites in Madagascar. These included another spectacular new find, a dinosaur with a whorl of protruding teeth at the front of its mouth, *Masiakosaurus knopfleri*, named after Mark Knopfler, whose music the field team loved. This was a beast so bizarre that its reconstruction made the cover of *Nature* magazine (Sampson *et al.* 2001). When I held this beautiful lower jaw, a 68-million-year-old bone that came from a large, lurking predator, the first thing I noticed was the absence of most of the teeth. Two teeth were still preserved in their sockets and the rest of the jaw was in pristine condition, so where were all the teeth?

Scott told me that they had a *National Geographic* film crew with them during the excavation of this dinosaur. It was late in the afternoon and the sun's light was rapidly fading, but they didn't want to take away all the bones until filming was complete. The plan was to wait until first light next morning, film the bones in situ, then remove the specimens and wrap them up. To their surprise, the next morning they found that the lower jaw, which only the night before had been resplendent with all its razor-sharp serrated teeth, was now almost toothless! All but two of the teeth had been forcefully removed from the jaw.

Who did this? Scott could only tell me that a certain French fossil dealer, who had had previous dealings in Madagascar, had been spotted in the nearby town when they were there. The most likely explanation is that somebody paid some of the locals to remove the teeth from the jaws. We can't point the finger at anyone in particular as having been behind this theft, but the desecration of a carefully excavated site, discovered by a professional team and about to be recorded on film for all posterity, is to me morally repugnant. Scott told me he wasn't too happy about it either at the time.

I asked Scott how the commercialisation of fossils had affected his regular palaeontological field work. He said that some of his old sites were becoming less accessible to work. Commercial dealers now paid landowners generous fees to allow them to search, fees the government institutions couldn't match, so they couldn't dig on the sites they had been working for years. In some cases commercial diggers are paying landowners for digging rights to newly discovered sites, thus denying scientists access to the sites. One example is that of a newly-discovered Cambrian site in southern Utah which is yielding beautiful fossil arthropods and soft-bodied fauna, similar in quality to the famous Burgess Shale site in British Columbia and Chengjiang in Yunnan, China. Unfortunately, because the land is privately owned, the site is now under the control of dealers and the material is being collected without scientists being able to study the site first hand. The potential loss of important scientific information here is enormous, as dealers search mainly for commercially valuable specimens, ignoring microfauna or insignificant-looking, scrappy fossils, which can sometimes be juveniles (larvae), representing the growth stages of the adult forms. Such specimens are invaluable in helping to determine evolutionary lineages through studies of growth variations in species.

Scott expressed his real concern about this situation to me, but shrugged his shoulders when I asked him what could be done about it. 'Nothing,' he said. The laws in Utah clearly favour the landowner above scientific endeavour. I thanked Scott for his time, and for lunch, and took a stroll through the beautiful campus of the University before heading back to the hotel.

I looked up the online catalogue of the dealers that were selling material from the southern Utah Cambrian site. I was amazed at what was on offer. For US$12 999 (as at March 2002) I could buy a complete specimen of the giant Cambrian arthropod *Anomalocaris*. This beast, first recognised from the Burgess Shale site, has been found recently in Australia, but there are very few relatively complete specimens. This being so, I believe that the sale of these specimens should be restricted by law until the full scientific significance of their site of origin has been properly documented. If this is not done, the context of the specimen within its stratigraphic framework, an intrinsic part of its value (to a scientist at least), remains unknown.

As I was finishing this book I received an email from Scott (27 March 2002) adding a final footnote to the situation in Utah.

Just found out yesterday that one of our new sites in southern Utah was just vandalised, with some bones stolen. This is a real tragedy since the site is producing a single disarticulated skeleton of a new ceratopsid and thus will be a Holotype specimen. The authorities think they know who it is and will try to go after him . . .

I was deeply saddened by Scott's email, but at least the perpetrators may be caught before they destroy too many more sites.

My interview with Scott would have to be my last for the time being, as my leave had run out. I headed back to Australia; Alan and the team were bound for South Dakota and some more interviews.

Fred and Candy Nuss love fossils. They seize any opportunity they can to get out and search for them, and look forward to finding that elusive big dinosaur skeleton. They have already found one partial *Tyrannosaurus rex* skeleton and the skeletons of two oviraptorosaurs, both on private land in South Dakota.

The team went out to the dig site with the Nusses. Driving through the eroding badlands of South Dakota, they could almost smell the dinosaur bones coming out of the gullies and creek beds. The Hell Creek Formation is a Late Cretaceous rock formation (around 65–67 million years old) that represents the most accurate record we have of the end of the reign of dinosaurs. It was a time when the giants, *Tyrannosaurus*, *Triceratops* and the mighty duck-billed hadrosaurs, roamed a landscape full of ferns, pines and early flowering plants, such as magnolias.

Fred and Candy both have day jobs, but since they began serious fossil prospecting they have found some very valuable specimens. Alan asked Fred why he loved fossil hunting.

I have a day or so of walking . . . approximately where I'm standing here, you can see how honeycombed the piece [bone] is, a sure sign of a Mr Rex or Mrs Rex. Everybody wants a *T. rex*. And, if we get lucky a little bit, we'll find where this piece comes from, and there'll be three or four more, which will lead us to ten more, which leads us to 50 more, and after a hundred bones we'll have

30 per cent of the animal. Everybody's looking and everyone wants to make a million bucks.'

Fred did find a partial *T. rex* skeleton some years ago. It made him reflect on his life, and he spends most of his time now as a professional dinosaur hunter. His biggest discovery to date is two oviraptorosaur skeletons.

Oviraptor was a small, toothless dinosaur first discovered in the 1920s by the Roy Chapman Andrews expeditions at the famous Gobi Desert sites in Mongolia. It had a crested head and its skull was about the size of a baseball. The first skeletons were found surrounded by broken dinosaur eggs, and the assumption was that these dinosaurs had probably stolen the eggs to eat them, hence the name *Oviraptor* (egg stealer). When the American Museum of Natural History expeditions led by Mark Norrell went back to Mongolia in the 1990s, they found more skeletons of *Oviraptor* associated with complete, undamaged eggs. It was then discovered that the dinosaurs were brooding over their own nests. These finds gave us one of the most remarkable insights into dinosaur behaviour.

Fred and Candy Nuss' skeletons were of a much bigger beast, something more than three times the size of the Mongolian species. On reflecting how one lucky find can change your life, Fred said:

A lot of people say 'Oh Fred, you're lucky' but, you know, you don't find them sitting at home or sitting around in a field or looking out the pickup, you just gotta do it. Sometimes you wonder why you're even doing it. Everybody else is in the modern world and I'm still digging in the dirt.

It gets under your blood and you just enjoy it. Sometimes you don't get anything but there's always next year, God willing.

It's just one afternoon over one hill, underneath one rock, can change it. Of course, if you get hit by a rattlesnake that can change it too.

The skeletons were of a giant *Oviraptor*-like dinosaur and the Nusses brought in their friend Mike Triebold, a fossil dealer based in Colorado, to help them prepare and market the specimens for sale. Mike put in countless hours of detailed preparation.

'This is a composite of the two skeletons that Fred found,' he said, standing next to the towering resin skeleton in his workshop. 'And fortunately what the first one didn't have the second one did have and they were almost exactly the same size.'

Mike didn't want to discuss the price put on the specimens. As they represent a new genus, the buyer will have naming rights to the new dinosaur. Alan asked Fred about the value of the specimens and how he felt about their being sold.

> We've had it priced at [US]$960,000—for the two *Oviraptor* skeletons and one reproduction, standing, and we would like these to be donated here in the United States, because of the good scientific value that they have, and to be studied here.
>
> There are certain things in the fossil business that you want to respect, the science. And those things are scientific, they're exciting. I mean people can learn something!

Mike Triebold had this to say about the significance of the oviraptor skeleton:

> From the scientific standpoint, it's important. And from the marketing standpoint it has great potential too. So both of these things in one skeleton is a real coup. And this has it. That's why it's the chicken from hell!

It's easy to see why Mike dubbed the specimen 'the chicken from hell'. It has very bird-like features, yet its large size and sharp curved claws definitely give it that *Nightmare on Elm Street* look. When I first saw it in Tucson, it made me drool. It would be a great specimen to study, one which could certainly tell us a lot of new things about dinosaur evolution.

Alan asked Mike Triebold what he wanted for the specimens.

It doesn't matter where the money comes from, whether it's a private donor to a museum or if it's a private purchaser of the specimen, or if it's a museum with their own patrons who purchase the specimen. The most important thing is they end up in a permanent repository where they would become part of the world's body of knowledge of dinosaurs. That's important to us.

A final word from Mike Triebold about the amount of time and effort that professional dinosaur dealers must put in to get a specimen ready for sale: Mike told Alan of a *Triceratops* skeleton that took some 15 000 man hours, from the time of its discovery in the ground, for the fully-prepared, mounted skeleton to be ready for sale.

'Well, put a pen to that and that's got to sell for a lot of money or you'll go broke in a hurry,' said Mike. In a later statement he said, 'Actually I did a calculation one time and on a typical elephant-sized dinosaur we make about as much as the local car repair shop.'

Most of us work a standard 37.5 hour week in Australia, which would mean investing eight years exclusively on one specimen that you then must sell to recover your investment of time and money in getting it ready for sale. At say a minimum professional salary of around AU$40 000 per year that's AU$320 000 before you've started to think about preparation materials and field costs (add another AU$140 000 for resins, vehicle hires, casual salaries, tools, metal framework and welding, and so on), so the specimen would have to sell for at least AU$460 000, just to break even. Add to this the value of all the time and effort expended in finding the thing, and the fact that you have a complete skeleton of a real dinosaur, and its value as a specimen becomes a complex issue.

A final reflection. When people ask me how much a fossil is worth, I sometimes ask them to reflect for a moment on the effort that went into collecting it. After twelve expeditions to the remote Kimberley district in Western Australia, I have found only one specimen of some species of fossil fishes. Add up the costs of twelve field trips to the Kimberley, vehicle hire, my salaried time and the preparation costs involved (time, chemicals and technical assistants' salaries), and that's your base cost for possibly getting a specimen of one of those rare species, if you are lucky.

Alan and the team were now heading for China, home of the world's largest black-market industry in fossils. It was risky business probing into the collection and illegal export of specimens, as it soon became clear that some powerful people were involved in fossil smuggling.

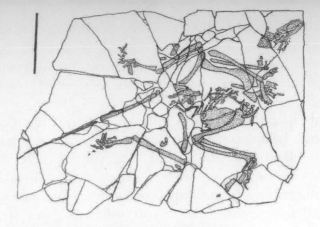

Dragon Bone Sale

9

Fossils of ancient vertebrate animals and ancient arthropods having scientific value receive the same state protection as cultural relics (19 November 1982, *People's Republic of China Cultural Relics Protection Law*, article two, section three).

Illegal managing (including the purchase, transport, reselling and profiteering) of cultural relics is a serious situation which constitutes a crime, and is punished according to the crime of speculation and profiteering. Illegal managing of grade three relics is punished by no less than 3 and no more than 10 years imprisonment, and can be punished by confiscation of property. In the case of 'Zhang Biliang and Others selling for Profit and Speculating on Fossilised Dinosaur eggs' (1995, Xixia County People's Court), the defendants had resold 156 dinosaur eggs, of which 148 were deemed to be cultural relics, grade three (extracted from Schmidt 2000).

Sketch of *Archaeoraptor*, a man-made composite containing parts of fossil birds and dinosaurs from China. (Bar scale is 10 cm)

To the Chinese, dragon bones are a valuable component of traditional medicine. The bones in question are the fossil bones of ancient large mammals, such as *Stegodon orientalis* and *Rhinoceros sinensis*, and are used as a sedative and tranquilliser for the treatment of palpitations, insomnia and dreamfulness due to neurasthenia and hypertension.

In 1903, Professor Max Schlosser of Munich published a treatise on dragons' bones entitled *Die fossilien Saugethiere Chinas*. According to Schlosser, these prehistoric animal fossils, known collectively in China as dragons' bones or dragons' teeth, included a variety of extinct species. Schlosser identified approximately 90 mammals whose bones were being sold as dragons' bones. Fossils had been used for centuries in China for their medicinal properties, but it wasn't until the 1929 discovery of Peking man (*Homo erectus*) at Zoukoudian Cave, near Beijing, by Chinese scientist Pei Wen-Zong, that the study of palaeontology really took off in China. The Peking Man fossils were kept in Beijing Concord Medical College until 1941. Several days before the outbreak of the Pacific War, to prevent their confiscation by the Japanese Army, the fossils were delivered to the American Army for their transfer to their wartime home in the American Natural History Museum. It was too late to protect the fossils, however. The train carrying the fossils was captured by the Japanese Army and the fossils haven't been seen since. Many thought they were captured by the Japanese Army and transferred to Japan, but this was strongly denied. Others thought the fossils had been transferred to the USA before the train was captured, or seized by the Americans during their occupation of Japan. The US government has denied both hypotheses. The whereabouts of Peking Man is one of palaeontology's greatest unsolved mysteries.

The first dinosaurs were found in China in the 1910s by Russians working in the border area along the Heilongjiang River. The first Chinese palaeontologist, Professor Yang

Zhongjian, better known to the western world as Professor C. C. Young, was educated in Germany. On his return to China in 1928 he set about searching for dinosaurs and other fossil reptiles, and made significant discoveries in Szechuan Province, including the gargantuan long-necked dinosaur *Mamenchisaurus*, an almost complete skeleton of which was unearthed in 1957. Today China has one of the world's largest palaeontological institutes, the Institute of Vertebrate Palaeontology and Palaeoanthropology in Beijing, founded in 1960. The Institute was formed primarily to house precious fossils of early humans, at the same time protecting and studying other vertebrate remains found throughout China.

China is home to many of the world's most significant fossil sites, from the remarkable Early Cambrian Chengjiang fauna of Yunnan through to superb Devonian fish sites in Yunnan and many spectacular dinosaur sites, especially in Szechuan and Liaoning Provinces. Since the Second World War palaeontology has become a major area of science in China, and nowadays the local governments in each of the provincial regions also recognise the enormous tourist potential of palaeontological museums. China has over 70 palaeontological displays in its regional museums. Every time an important new fossil site is discovered, a new museum is built to house the collection and attract local and overseas tourists to the region.

I had briefed Alan Carter and the film crew about the extent of the illegal fossil trade in China. The purpose of their visit was twofold: first, to find out about the black market in Chinese fossils by interviewing scientists at the Institute of Vertebrate Palaeontology and Palaeoanthropology in Beijing, and second, to visit sites where illegal fossils had come from to see how well, if at all, these sites are protected. They also planned to travel to Jinzhou and meet with Mr Du Wenya, a local fossil collector who had amassed a very large collection of rare and valuable Chinese and overseas fossils.

Protecting fossil sites in China

China has always valued its heritage, yet measures to protect its fossil heritage have only been in place since the early 1960s. I asked Professor Zhu Min, current Director of the Institute of Vertebrate Palaeontology and Palaeoanthropology, for his views on Chinese laws pertaining to fossil protection. He replied (by email on 9 April 2002):

> First, I would like to point out that all Chinese vertebrate fossils for sale in the international fossil markets are illegal according to the Chinese law. The so-called 'cultural exchange documents' to export Chinese vertebrate fossils are also illegal. There is no legal ground to export Chinese vertebrate fossils, in other words, all the Chinese vertebrate fossils for sale are the smuggled fossils, and Chinese government has the rights to ask for the return of the fossils, as we did with *Archaeoraptor*.
>
> The Chinese Cultural Relics Protection Law was issued on Nov. 19, 1982. Before this law, the government issued various regulations to protect Chinese cultural relics including vertebrate fossils. In 1961, the Chinese government issued Cultural Relics Protection Provisional Regulations (or Rules).

Today, Chinese officials deal strictly with any fossil smugglers who are caught red-handed. For example, in August 1995 three Chinese men were caught trying to sell dinosaur eggs. Zhuang Weimin, 60, bought sixteen dinosaur eggs in Henan province and, together with his accomplices, arranged their sale in Shanghai for 450 000 yuan (US$54 216). The three were arrested as the eggs were being handed over. Zhuang was found guilty of violating laws protecting Chinese cultural relics and jailed for five-and-a-half years. The other two were each jailed for five years (*Xinmin Evening News*, 23 June 1995). The case mentioned at the start of this chapter, that of Zhang Biliang, Liu Dezhi and Zhang Chunling, who were caught with 156 dinosaur eggs they bought from local collectors in Henan Province, also resulted in severe jail sentences. Zhang Biliang and Liu Dezhi each received six years, Zhang Chunling four years. Despite such heavy penalties for people caught trading in fossils, the trade in illegal Chinese fossils is openly conducted around the world. Almost every rock and fossil shop I know of in Australia and elsewhere sells Chinese dinosaur eggs and many international dealers openly advertise them in their online catalogues, along with Chinese dinosaurs and fossil birds. As I have already explained, those selling the fossils are breaking no law, although unless these fossils were accompanied by export papers confirming that they were legally exported from China as cultural exchange items, they are sure to have been smuggled out of the country. Most fossil dealers purchase the Chinese fossil eggs directly from the dealers who sell them at trade shows like the Tucson show.

The world's museums are well aware of the illegitimate trade in Chinese vertebrate fossils, and some of those which have been offered rare and scientifically important Chinese fossils have deliberately shied away from them. The commonest Chinese items on the international black market are dinosaur eggs from Henan and Hubei Provinces, and fossil fishes, birds, dinosaurs and marine reptiles from Liaoning Province. In an effort to protect dinosaur egg beds in Hubei, provincial officials have set up a 15 sq. km protection zone around the eggs. A similar area exists to protect fossilised eggs in Guangdong Province and all construction projects in the area must be approved by the Cultural Relics Protection Project.

In Liaoning, the fossil beds have been designated as a 'Fossil Birds Preservation Zone' ('the Zone'), a 46 sq. km region south of the town of Beipiao. The Chinese equivalent of over a million US dollars has been pumped into the region for fossil protection measures. This includes the building of a new fossil museum in Beipiao, ten kilometres south of the township.

'Here they are building a tourist region so we've built a museum on the fossil site and maybe a few people will visit,' said Professor Ji Qiang. Not all Beijing scientists, however, are happy about such rare fossils being stored in small provincial museums.

'For the moment I think the situation is very, very, very bad,' said Dr Zhonghe Zhou. 'I mean those so-called museums, I wouldn't call them museums I would call them a warehouse. They don't have curators. I don't think they can be very well curated. No preparation, no study. But the curation concerns us most.'

A town official, Zhao Yibing, has been named administrator of the Zone. He employs five full-time guards who provide round-the-clock protection against thieves. However, this scheme depends on the honesty of the guards, their ability to police a large area and the availability of the necessary funds.

Alan and the crew drove out to see the fossil site at Beipiao, where Alan interviewed Zhao on his strategies for combating fossil thieves.

> The building up there is the guard's house. There are people on guard 24 hours a day protecting the fossils. Also the scientists visit the local people and educate them in the importance of fossils and why they need to protect them.
>
> Since Liaoning introduced its legislation for the control of fossils we have enforced the laws diligently whilst also carrying out a lot of public education. As a result the illegal trade in fossils has been eradicated.

In Liaoning the fossil protection agency is also responsible for hiring diggers and regulating how much material can be sold (the less important specimens, such as fish or plants).

'It's definitely very bad to have the Protection Agency collecting fossils,' said Dr Xu Xing. 'It's like in China you have those government officials opening a company of the

Court to have a company of their own. Can you judge a case involving your own company or business?'

'Last year there were too many fossils circulating in the market,' said one official at the Fossil Protection Agency, 'because there was too much digging. This year we employed three more people to strengthen fossil protection. Now there aren't many [fossils] out there. It's hard to find any good ones in the market.'

In effect, any Liaoning specimen in private hands anywhere outside China was illegally obtained. Bearing this fact in mind, here are some of the recent controversial cases involving Liaoning fossils sold on the international market.

Liaoning, a hot spot for fossil smuggling

The discovery in 1991 of the famous Liaoning sites of the Yixian Formation, marked the beginning of a series of important palaeontological discoveries that would change our views on the evolution of birds and the biology of dinosaurs. The sites yielded a series of extremely well-preserved early birds, with feather patterns intact (see Ackerman 1998), which bridged the longstanding hiatus in avian evolution from the oldest bird, the German *Archaeopteryx*, to modern birds.

In 1996, a paper published in *Nature* introduced to the world one of the best-preserved dinosaur fossils ever found—the complete skeleton of a small predatory dinosaur, a coelurosaur named *Sinosauropteryx*. Not only was the body covered in a fine feather-like integument (the first direct evidence of such structures on a dinosaur), but it had eggs in its oviduct and the remains of a mammal in its gut! Even more remarkable have been the discoveries coming out of the sites around Beipiao. To date, more than five different fossil birds, and seven varieties of dinosaurs exhibiting feathers or protofeather-like integumentary structures have been found. These finds have not only closed nearly all the gaps in the evolution of dinosaurs into birds, but have given us detailed insights into the early evolution of true birds from feathered dinosaurs, even showing us how they became able to fly. These sites are hugely significant scientifically, as they keep yielding more and more spectacular dinosaur finds, such as the recent discovery of a perfectly-preserved, complete skeleton of a dromaeosaur or 'raptor' with quite complex feathers adorning its arms and a scattering of fine protofeathers all over its body (Norell *et al.*).

The sites have yielded such a high number of significant fossils that worldwide attention has focused on their monetary value. After the discovery of the first fossil bird from Liaoning, it wasn't long before specimens started to appear on the market (Hecht 1996). Despite the moral dilemma of buying illegally exported fossils, specimens were snapped up by several major European and Japanese museums. When asked about the legality of buying Chinese fossils from Liaoning. Professor Ji Qiang, Director of the National Geological Museum in Beijing, was blunt: 'There is no such thing as a fossil from Liaoning which was obtained legally'.

Alan and the team spent a day at Beipiao watching the local farmers, working under the direction of scientists, carefully excavate the greyish-green shaley layers. The men carefully removed overburden, then cleaned the flat surface of the rock, chipping away at the side until large slabs of flat rock can be removed. These are gently split, in the hope that the fracture will find its weakest path through a good fossil. The labourers aren't paid well (US$2.40 per day), but all the specimens from this site will go back to the Institute of Vertebrate Palaeontology and Palaeoanthropology, in Beijing.

After the team had finished filming the dig sites and local scenery, the scientific supervisor packed up for the day and left. Alan and the crew were still at the site, with their driver and guide, and it was then, Alan told me, that he saw one of the local diggers come up to another man and show him a large fossil fish. The two men had an animated conversation, regardless of the fact that they were being filmed. Later Alan had the conversation translated from Chinese, and this is an exact transcript.

1st man: 'I'll prepare it when I get home.'
2nd man: 'Do you think we're going to be in trouble?'
1st man: 'I don't think so.'
2nd man: 'Be careful, the police might catch you later. You're going to be top of the "most wanted" list. They've even got a photo of your numberplates.'

Archaeoraptor

In 1993, a Chinese businessman came to Phillips Fine Art Auctioneers in New York with a clutch of fossils he had bought from farmers in China. Scientific consultants who checked the collection spotted a superbly preserved specimen of a primitive beaked bird that was unlike anything they had seen before. Phillips will not sell fossils that are 'undescribed or new to science', and David Herskowitz, the then organiser of the company's annual natural history auction, says the consultants 'advised him not to touch the bird'. *Confuciusornis* was thus first seen in the West by a fossil dealer rather than an academic at a museum or university.

The genus *Confuciusornis* was only formally published in 1995 by Professor Hou and his colleagues. Since then, scientists in China have found many other specimens (see Chapter 7). Now that it was known to science and, according to the auctioneers, well-represented in museum collections, Phillips decided to put its fossil bird up for auction, despite the current Chinese laws prohibiting legal export of the fossil birds (Hecht 1996).

The case of *Archaeoraptor liaoningensis* is the most notorious case of a smuggled fossil in years. The fossil, touted as a rare missing link between birds and dinosaurs, was sold at the Tucson fossil show in 1999, and later shown to be not what it seemed. Apparently a farmer in Xiasangjiazi, in Liaoning Province, had found a fossil bird but the slab split into many pieces. Nearby were the remains of a small dinosaur, its tail intact on a separate piece of rock. Back at his house he tried to rejoin all the pieces (making a very skilful job of it), and fitted the tail of the dinosaur to the body of the bird, whose own tail was missing. The resulting fossil was every palaeontologist's dream, a mysterious creature with a very bird-like body, wings with flight feathers and a typical primitive reptilian tail. Only in the famous *Archaeopteryx* had such features been seen before.

The specimen was sold to a dealer who had contacts in a scientific institute in Guilin. His accomplice was able to get him a legal document allowing export of the specimen, as part of 'a scientific exchange program'. All document-ation was, of course, faked, possibly arranged by large amounts of money changing hands, as such a specimen, showing intermediate characteristics between dinosaurs and birds, would never have been allowed out of the country.

The fossil was bought by Stephen Czerkas, of the Blanding Dinosaur Museum in Utah, who was amazed by it when he first saw it in a Tucson motel room, and managed to find financial backers to help him meet the US$80 000 price tag. The name *'Archaeoraptor liaoningensis'* was given to the specimen in the *National Geographic* magazine (Sloan 1999). Czerkas then coopted fellow dinosaur expert Phil Currie of the Tyrrell Museum in Canada as coauthor of the description of the new specimen, but only after Currie had convinced Czerkas that the specimen must be returned to China after any paper was published so as not to upset future working relation-ships with Chinese palaeontologists. Currie also argued that a Chinese scientist from the Institute of Vertebrate

Palaeontology and Palaeoanthropology, Xu Xing, should be invited to the USA to spend three months there as part of the team studying the new specimen. Czerkas agreed, and when Xu was brought over the team of scientists got busy looking at their prize specimen.

To cut a long story short, the specimen was soon suspected to be a fake, a composite of two animals, by virtue of its skeletal anomalies, and this was confirmed by the US$10 000 worth of CT scanning done by Professor Tim Rowe at the University of Austin, Texas, which revealed that 'the tail had no natural connection to the body' (Rowe *et al*. 2001). In fact, the specimen was shown to have been a mixture of up to five different specimens that together had formed the world's greatest biological mosaic.

This story does have a happy ending. The specimen was handed back to China on 25 May 2000. Xu Xing undertook further investigations in China and managed to track down the counterslab, with the other side of the tail *in situ*. Xu still believes that although *Archaeoraptor liaoningensis* is a mixture of different animals, its various parts could be very important specimens, once identified. More parts of it were found at the original site and he recently described it as a new dinosaur, which he named *Microraptor* (Xu *et al*. 2001).

Lewis Simons, writing for *National Geographic* magazine (Simons 2000), visited the site in 2000 and was told by a local police official that only farmers authorised by the police can dig for fossils, and every fossil they find must be turned over to the authorities. A judge from Jinzhou told him that punishments for not surrendering fossils range from 2–3 years' jail to, in cases where the fossil has been smuggled out of the country and sold for large amounts overseas, execution.

An exquisite Liaoning dinosaur
One of the most remarkable dinosaur fossils to come out of Liaoning in recent years is the small skeleton of a

1 The stolen dinosaur prints from Broome, Western Australia.

2 John Yates (left), Steve Rogers and me at Broome police station examining the
   stolen dinosaur footprint that was recovered in 1999.

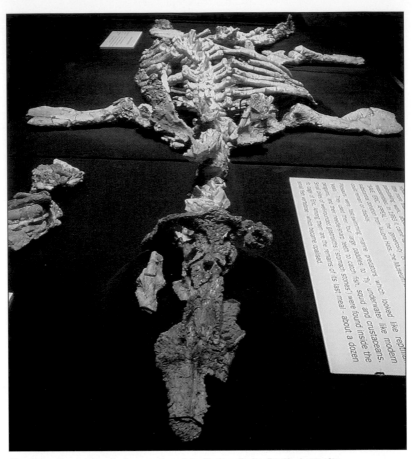

3   Eric, the opalised pliosaur found in Coober Pedy, South Australia.

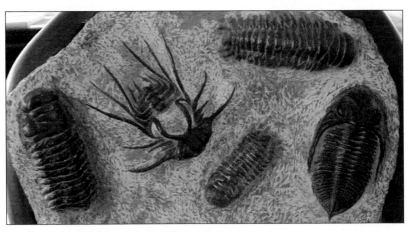

4   Moroccan trilobites for sale—some faked.

5   Bob Bakker (with hat) and me (lying down) digging up dinosaurs at Como Bluff, Wyoming.

6   Aerial photo showing poaching sites in Wyoming, where fossils have been taken from government lands.

7   Steve Rogers with his plane. Steve flies aerial patrols over State lands in Wyoming, searching for fossil poachers.

8   *Phaerodus*, a Green River (Wyoming) fish.

9   Mike Treibold and the 'chicken from hell', a new genus of oviraptorosaur
     dinosaur from South Dakota.

10 Carl and Shirley Ulrich, who have been quarrying the Green River fish sites for over 50 years. In the background, a perfectly prepared palm frond with fish.

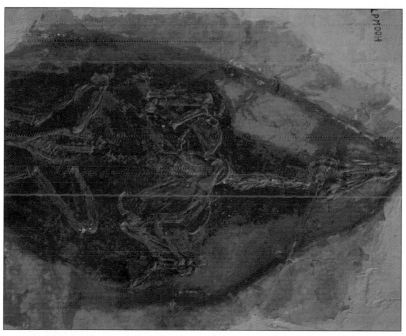

11 *Confuciusornis*, an early fossil bird from Liaoning, China, on display in Scipio Fossil Museum.

12 Workers searching for fossils at Chaoyang site in Liaoning Province, China.

13 Dinosaur eggs from China, one of the items most commonly smuggled out of the country. These specimens are safe in a museum in Jinzhou, China.

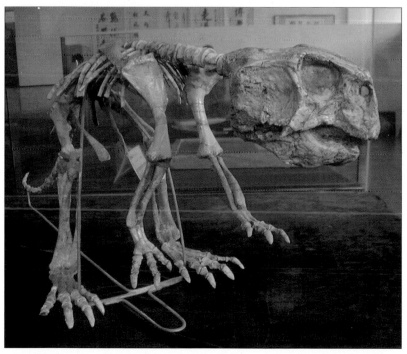

14 *Psittacosaurus* skeleton from China, the most commonly smuggled and sold type of dinosaur skeleton in the world.

15 A Brazilian *Mesosaurus* for sale in Australia.

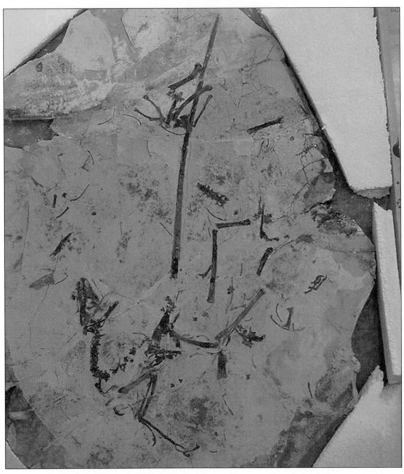

16 A possible new type of dromaeosaur (raptor) from Liaoning, China, on sale at a Tuscon show in 2002 for US$68,000.

psittacosaurid, a plant-eating dinosaur that is one of the earliest ceratopsians. It has the same parrot-like face, but lacks the frills and horns of the later ceratopsian dinosaurs, the most well-known of which is *Triceratops*. Several *Psittacosaurus* specimens have been found at Liaoning, in a slightly different sedimentary layer from the other feathered dinosaurs. This new specimen is exquisite in that it shows quill-like structures preserved on the rump of the animal, the first example of an ornithischian, or bird-hipped, dinosaur with such integumentary structures preserved. As such the specimen is extremely important in reassessing our ideas about dinosaurs and their appearance.

This specimen first appeared at the Tucson show about three years ago. The American Museum of Natural History's dinosaur expert, Dr Mark Norell, saw the specimen and wanted the museum to purchase it. Due to its close relationship with institutions in China, however, the American Museum of Natural History did not want to be seen to be involved with any fossils which were obviously smuggled out of China. Dr Norell's comments on the specimens were: 'This specimen could redefine how we look at dinosaurs, but we can't say, since no one has studied and published on it. It should be returned to China, where it can be studied in a museum' (Dalton 2001b).

The specimen was then in the hands of a German dealer who had had it prepared by a dealer in Italy, Flavio Bacchia. Bacchia had shown the specimen to palaeontologists at the Natural History Museum in Milan, and apparently attempts were made to involve Chinese palaeontologists in co-operative research on the specimen. Perhaps this attempt at international liaison failed because the specimen had been smuggled out of China.

The specimen was also examined by Dr Eric Buffetaut, a dinosaur expert at the University of Paris. Dr Buffetaut had extensive professional ties in Southeast Asia and China, and he made efforts to have the specimen repatriated to China, but without success. The follow-up to this story was published in late 2001 (Dalton 2001b). The specimen was snapped up by the Senckenberg Museum in Frankfurt, from a German dealer, Ulrich Leonhardt, for around US$200 000, which the Senckenberg is paying by instalments. When the museum's Director, Professor Steininger, was asked about the legality of the specimen he said that the museum had proper German importation documents and exportation records from the USA, where Leonhardt had apparently bought the specimen. But he would not discuss whether the museum had any paperwork approving the export of the specimen from China.

Recently the case of the Liaoning psittacosaurid took another interesting turn. The following report is taken from the Chinese Xinhua news agency's report of 16 January 2002.

The Chinese Society of Vertebrate Palaeontology (CSVP) recently sent an open letter to Naturmuseum Senckenberg in Frankfurt, Germany, appealing for the return of a psittacosaurid fossil smuggled out of China.

The open letter reads that the museum has violated Chinese laws and international conventions by buying a series of fossils smuggled from China, including a psittacosaurid fossil bought at $200 000 U.S. last summer and several *Confuciusornis*, a famous Mesozoic bird discovered in Northeast China.

'Buying smuggled fossils will only jeopardise research and further propel more illicit collecting and underground trading of precious fossils,' said Dr Zhou Zhonghe, researcher with the Institute of Vertebrate Palaeontology and Palaeoanthropology of Chinese Academy of Sciences.

The rare psittacosaurid fossil attracted worldwide attention last summer for the odd integument structure of the tail, which will provide a new view on the appearance of dinosaurs, said Xu Xing, a dinosaur expert in the same institute as Zhou.

Xu added that researchers aware of international conventions refuse to study smuggled fossils and prestigious periodicals including *Nature* and *Science* forbid the publication of any research on smuggled fossils.

The fossils were most likely smuggled from the Western Liaoning Province in Northeast China, where in recent years there have been many excavations of rare fossils, as well as a flood of illegal digging.

Unscientific digging, fuelled by the smuggling and collection of illegal fossils, has caused a loss of information about locality and stratigraphy, essential information for scientists researching this area.

A local paper, *Chaoyang Daily*, reported on January 8 that there are hundreds of people digging for fossils every day in Dapingfang Town in Chaoyang County and hillsides are covered with ravines and big holes dug by peasants.

As one of the most leading natural history museums in Europe, Naturmuseum Senckenberg has a responsibility to educate the public to fight illegal fossil trading and Chinese experts hope serious consideration will be given to the return of the fossils, the open letter stressed.

We can only wait and see how the Senckenberg Museum responds to this demand. Although they may have a legal case for having their money reimbursed if they decide to return the fossil, their actions will have much wider ramifications throughout the museum fraternity. If the Senckenberg did decide to repatriate the psittacosaurid fossil it would be obligated (as specifically requested by the Chinese letter) to return its other Liaoning fossils, which include a number of well-preserved fossil fish, birds and small reptiles. To make the money available to reimburse the museum might set a precedent that only if funds were available should other museums return their fossils to China. As there are now a fair number of foreign museums holding Liaoning specimens, the cost to China of buying back her stolen heritage would be enormous.

Professor Zhang Meeman is one of China's most respected palaeontologists, and a previous Director of the Institute of Vertebrate Palaeontology and Palaeoanthropology. Alan asked her how she felt about the illegal fossil collectors who raid some of the Liaoning sites to steal fossils.

> Everything was taken away, not by specialists [or] experts, but by people in order to make money and this is, I mean, it's rather encouraged than controlled, if I [may] say frankly.
>
> Many times they break the fossils and they don't know which is important, which is less important. They just collect the complete specimens. Sometimes the complete specimens are less important than the fragmentary ones.

When asked whether she thought the laws in China were doing anything to put a stop to the illicit fossil trade, Professor Zhang replied, '. . . The laws I think do not play any importance at the moment. It's a great pity.'

Chengjiang fauna for sale

The Chengjiang fauna is perhaps one of the world's most interesting windows onto the explosion of life at the Cambrian boundary, 520 million years ago. The material comes from a site in Yunnan, southern China. Since its discovery in 1984 by Hou Xianguang, more than 70 species of exquisitely preserved soft-bodied fossils have come from this remarkable site. In more recent years, the world's oldest fossil fishes and boneless vertebrates have been identified from Chengjiang, along with an entirely new phylum of animals, the Vetulicolia (Shu *et al.* 2001). As such the site's significance to palaeontology, to our understanding of the very origins of modern animal phylogeny, is incalculable.

Chengjiang fossils have not been traded or sold to any legitimate museum. Some museums, such as the Gamagori Museum near Nagoya, in Japan, working closely with Chinese scientists, have negotiated official loans of Chengjiang specimens for display. None has obtained specimens for its own collections. In China the material is considered far too valuable scientifically to be made available commercially. Despite this, however, Chengjiang fossils can now be bought over the Internet. A site based in North America offers a wide range of Chengjiang fossils with a total sale value of around US$10 000. The same site also sells unique material from the Cambrian of Utah, including a rare, complete specimen of the early large predatory arthropod *Anomalocaris*. One can only presume that since the significance of the Chengjiang fauna has been publicised in scientific papers, locals have been getting in to the site and searching for specimens to sell on the black market. It would seem that the same sort of fossil rustling as is going on in Liaoning is also going on in Yunnang.

Jinzhou

Alan and the team moved on to Jinzhou to meet with Du Wenya, arguably China's largest private fossil collector. In January 2000, he opened the first private fossil museum in China, the Jinzhou Wenya Museum, to display his collection of more than 10 000 specimens. The museum was set up on the proposal of Professor Hou Lianhai, a noted palaeontologist who specialises in fossil birds. Du's collection contains the first specimens of *Confuciusornis*, and the only known specimen of *Confuciusornis dui*, named in his honour, is also on display.

Du has a large collection of local fossils from Liaoning Province, including large fishes, dinosaurs (*Psittacosaurus* and *Sinosauropteryx*), marine reptiles (*Sinohydrosaurus* and *Kuiechosaurus*), ammonites and trilobites from Morocco and many other attractive items. He works part-time as a city prosecutor, and earns some of his income from what he calls 'box office' (visitors to his museum) and from the private sale of common fossils. In recent years he has taken his collection on tour, most recently to Chicago in December

2000. Du has been collecting fossils from the Liaoning region ever since the sites were discovered. As his collections are open to the public and accessible to Chinese academic institutions, his museum is seen as a great asset to Jinzhou and a good way to promote tourism to the region.

When Alan asked him about a particularly nice dinosaur skeleton he had on display from Liaoning, and the legal issues surrounding his fossil collection, he answered:

> This is a dinosaur. This is the largest discovered in Liao Xi. It's two metres long. We bought the specimen for 50 000 yuan, but now it would cost around 200 000 yuan.
>
> These were collected in the market place in Liao Xi at a time when there were no regulations. There aren't many collectors and most specimens were heading for overseas sale. Now there are no more. Even if you offered millions, you wouldn't find a good specimen. Basically they are gone.

Alan and the film crew then visited a fossil market in Jinzhou. Outdoors, in the open air, there were many people selling the local fish and plant fossils, all quite legal as these are not deemed to be 'cultural relics' on the same level as the birds or dinosaurs. He enquired if there were any dinosaurs for sale. A Chinese woman took them back to her house, where she unwrapped a small, almost complete dinosaur skeleton on the floor in front of him. It was a *Psittacosaurus*. The conversation, as recorded, went like this:

> 'So what is this?' Alan asked her.
> 'It is a species of dinosaur,' she said.
> 'So how much is this?' asked Alan.
> '5000 yuan,' she replied.
> 'For me it would be difficult to get this, to take this home.' Alan explained, turning to the translator. 'Does she have any ideas how I should do this?'

'Yes, don't pack all the bones together. Pack some in one bag and some in another, then they won't be able to identify them.'

'The bone is very strong,' she said. 'This is the nose. This thing is rare. We bought quite a few. Pack the bones into separate bags, then they won't be able to identify it. This is how you take it out.'

'If you want to take something away, take this one,' said the woman.

'This is so easy to take out. We have a friend who brought one of these out to the USA two days ago. He said it sold for $400 US.'

Warren Somerville, whom we met in Chapter 4, has a *Confuciusornis* in his collection. He told Alan how he had obtained it.

I had to unwrap it a few times to show people what it was. I paid $10 US stamp duty which went straight into the custom man's pocket. There's a lot of that and that's why the material comes out. It's a little harder for them to persuade them for a container load and there's a lot of these coming out of China too.

Warren simply bought the specimen in China and openly and honestly declared it at customs, paid his stamp duty and walked out with it.

In light of the large number of significant fossils discovered in recent years and Chinese legislation's failure to protect them, the National People's Congress is in the process of amending the *Cultural Relics Protection Law* to better protect fossils. The case of the Henan dinosaur eggs and the problems of protecting the Liaoning birds have prompted Chinese scholars from the Chinese Academy of Sciences to appeal to the government for greater legal protection to prevent improper excavations. Their recommendations include that certain departments and research institutes jointly control the issue of digging permits and designate a committee of the Ministry of Science and Technology to co-ordinate fossil searches and excavations (Schmidt 2000). The only thing the West can do to help is not condone the sale of any Chinese vertebrate fossils.

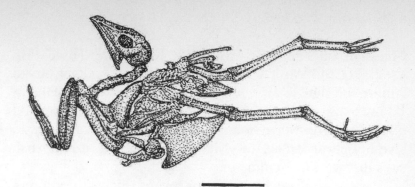

The Fossil Fish Capital of the World

10

Over the years Wyoming's Green River Formation has yielded the world's most significant Eocene terrestrial fauna—insects, fishes, reptiles, birds and bats—all magnificently preserved. It is also the world's richest fossil deposit, whose bounty has been harvested for over a hundred years and whose specimens are sold to every rock and mineral trade store in the world. Yet despite this, there are still troubles with poachers and illegal diggers. Where there's money involved, there's crime.

Sketch of *Gallinuloides wyomingiensis*, one of the very rare Eocene fossil birds from Wyoming. (Bar scale is 10 cm)

The sign outside the town read 'Kemmerer, Fossil Fish Capital of the World'. I had to stop the car and get out to admire the sign, as after all, I am a fossil fish expert. Finally, I'd made my pilgrimage to the world's fossil fish Mecca. I'd offered to drive Steve home when we left Salt Lake City, to give myself an excuse to visit the famous Green River fish sites just west of Kemmerer.

The Green River Formation is a very thick succession of sedimentary rocks representing three ancient lake systems, which existed from about 57 million years ago (the Paleocene Period) through to about 38 million years ago (the late Eocene Period). Most of the fish fossils come from the Fossil Lake Deposit, and these fish are about 49–53 million years old. Other fishes and animals from the Lake Gosiut Deposits are a little younger, dated at around 46–48 million years old. The Lake Unita Deposits don't yield as many fishes but are famous for their fossil mammals, and are well exposed in neighbouring Utah.

Fossil fishes were first discovered in the Kemmerer region back in 1856 by Dr John Evans, a geologist. He sent his first specimens to a well-known palaeontologist, Joseph Leidy, who named it as *Clupea humilis*, but the name was later changed to *Knightia eocenica*. *Knightia* is the most abundant fish fossil in the formation and has been made the official State fossil emblem of Wyoming. By the 1860s large quantities of the fossils were being collected, as railway workers uncovered the rich fossil fish layer and gathered specimens for State geologist Ferdinand Hayden. Hayden gave the fish to palaeontologist Edward Drinker Cope, who eventually went back himself to collect more fishes and write scientific papers describing them.

Since the 1900s a small number of collectors have been responsible for supplying almost all of the world's private and museum collections of Green River fishes. Robert Lee Craig collected fish from the region for over 30 years, starting in 1897. David Haddenham joined Craig in 1918 and, with his sons and grandson, worked the site until about 1970. Carl and Shirley Ulrich began working the region in about 1947. Robert Tynsky and his family began quarrying the fishes in 1970 and, like the Ulrichs, still work there today (Grande 1984).

After dropping Steve in Kemmerer, I drove another 20 km or so to the town of Fossil, Wyoming (with a population of two, according to the rusted sign). The unusual, rounded wooden house that dominates the town is the home of Carl and Shirley Ulrich, who have been working the Green River Shales for fossils for over 50 years. It is situated just opposite the spectacular Fossil Butte National Monument, now a protected site policed by pistol-packing palaeontologist and park ranger Vince Santucci.

On entering the Ulrich house I introduced myself to Shirley, who was minding the fossil gallery. The rest of the film crew had met Carl and Shirley a few days earlier, when they had flown to Kemmerer with Steve and called in at the gallery. Shirley was pleased to meet me, and told me how much she liked having Australians working up at their quarry. (So much so that she extended an invitation to me to tell anyone I knew back home that there was regular work there if they wanted to come over and dig the sites during summer.) The gallery was full of amazing specimens, each one a unique work of nature's art, the essence of what was once a living organism. The fish are almost perfect, their dark, complex skeletons contrasting well with the light, greyish-yellow shales. Some of them are enormous. A large garfish, about 1.5 metres long, formed the gallery's stunning centrepiece. The beauty of Green River fishes, though, is brought out largely by the skill of the preparator. Each preparator, like any other artist, has his own methods and unique skills. Fossil fish aficionados can even identify individual preparators' work.

In the downstairs workshop area Carl Ulrich, swathed in a huge apron, and with magnifying spectacles over his eyes, was intent on his workbench; one hand holding the buzzing drill tip, the other stretched out over a large slab of shale covered in fish fossils. Some of these fishes were cleanly prepped whole skeletons, others were still struggling to emerge from the rock.

'At times it's been hard to scratch a living from the fossils,' he told me in his slow drawl, 'but it's a job I have greatly enjoyed over the years.'

Carl and Shirley are model examples of people who work in the fossil-selling industry. They lease a quarry and, with the help of family and enlisted workers, open it up each summer to quarry out the rich eighteen-inch layer of fossils. Occasionally rare things turn up, such as remains of fossil birds, but more than 99 per cent of what is excavated is run-of-the-mill fish, plants and small insects. All of this material can be prepared and sold without government intervention. In the State of Wyoming, the Board of Land Commissioners lays down the rules and regulations for commercial and scientific permitting. Common invertebrates and five common species of fish can be quarried and sold without review by the Wyoming Geological Survey and without payment of royalties. Rare species such as fossil garfish, rays, bowfins and paddlefish can be sold without review, but the State requires that they be reported and that royalties be paid. All rare and unusual specimens—birds, crocodiles, turtles, lizards, snakes or bats—must be presented to the Office of State Lands and Investments within 30 days of discovery for review by the Wyoming Geological Survey.

Carl was busy preparing a huge slab of rock, some three metres long, covered in a swarm of small fish fossils (*Knightia*). He had been working on it for weeks, between other jobs. He took me around his workshop, showing me many examples of his fine craft work, from tiny insects and large plants to many different kinds of fishes, including the rare examples that fetch high prices; fishes that have eaten other fishes, or two specimens juxtaposed in an unusual pattern. He then pulled out the remains of a small fossil bird.

'This is one I'll have to hand in to the authorities,' he told me. 'Bird fossils are very rare here.'

Another huge slab containing a perfect large palm frond and single small fish was his latest masterpiece, now almost finished. One small slip with the electric vibratool or drill, however, and the whole thing could be ruined. Unlike a master painter who can rework his brush strokes, or erase or cover up material to make the final image perfect, the fossil preparator has only one chance to extricate the perfect form of the fossil from the rock. Carl truly is a living master of his art. Every fossil prepared and sold by the Ulrichs is signed by Carl, his distinctive signature etched into the corner of each slab. The Ulrichs also issue a certificate of authentication with each piece. Many of the finest fish specimens are tastefully framed and sold as works ready to hang.

When I told Carl and Shirley I was a fish palaeontologist, they kindly offered to take me out to their quarry. One of their assistants took me in their four-wheel-drive, up a very steep slope to the top of a nearby white hill of flat, layered sediments, where the famous eighteen-inch layer was exposed. There, on a perfectly flat slab of shale, I could see the rippled layers of sediments in the shapes of little fishes. The fishes are detected by their outlines, before the slabs are quarried out and taken down to the lab for the delicate job of preparation. There are around a dozen or so quarries leased in the region immediately adjacent to Fossil Butte National Monument. Some of these quarries are on private land, others are on State land that is leased for the commercial mining of the fossil resources.

Rick Hebdon, who runs Warfield Fossils, allows visitors to come and dig up their own fossils on his private quarry for about US$35 a day. Hebdon believes that palaeontologists working in museums need to be able to raise money to buy fossils from the private sector. He has an extensive collection of fossil birds that once caught the eye of a palaeontologist from the Smithsonian Institute, but the museum wasn't able to raise the US$80 000 asking price for the specimens.

Although the fossil-quarrying industry in Wyoming seems well regulated, it hasn't always been this way, and collectors who poach fossils from government lands are still a major headache for law enforcers like Steve.

Operation Rockfish

Steve first became aware of fossil poaching in the early 1990s when, while flying routine patrols, he noticed a large number of strange holes on State land. He requested that agents from the Bureau of Land Management and the Federal Bureau of Investigation accompany him on an aerial survey.

'At first we didn't have any idea what the holes were all about. Our best guess was illegal dumping of toxic chemicals,' he said.

Closer inspection revealed that these were sites where unauthorised collectors had dug down to the layers containing fish fossils and taken as much out as they could, leaving the holes unfilled. From the air it looked like a series of little bomb craters, unmistakeable evidence of the land being raped for profit. In 1992 Steve was instrumental in forming Operation Rockfish to combat the poaching of fossils from government lands in Wyoming.

When Steve started to investigate fossil smuggling, he soon found out that some players in the illegal fossil trade were closely tied to the illegal drugs and weapons trade. What was initially thought of as a local problem expanded to global proportions with Wyoming fossils being traced to Asia, Europe, South America and the Middle East, and these were only the Wyoming fossils. It soon became apparent that this was far too big for the Sheriff's Office to

The Fossil Fish Capital of the World

handle, and that a multi-agency effort was required. Steve recruited support from the National Park Service, the Bureau of Land Management, the US Forest Service, the National Guard, the FBI, the FAA and several other high-level State and Federal agencies. He soon found himself co-ordinating a huge, high-profile operation.

'It wasn't as bad as it sounds,' he remembers. 'Whenever we asked somebody to help us, we made it clear that this was a team effort with a common goal. If anybody's personal or agency ego conflicted with that, they didn't get invited back. We tried to conduct all of our planning and brainstorming in a very positive, team approach.'

To avoid potential inter-agency rivalries and conflicts, when it came time to divide the task force into specialised task-specific teams, Steve made sure each team had at least one representative from each participating agency. This also meant that each agency shared in all the available information. Routine aerial patrols enabled Steve to spot poachers in the act and detect sites where poachers had been active, so that ground patrols could then intercept the offenders. As the poachers cottoned on to the regular air patrols, however, they started working the illegal sites at night. Steve would notice new poach sites every morning. This prompted him to fly night patrols, to spot the poachers working by torch and gaslight. Next the poachers tried working under canvas covers, so as not to reveal their location. Steve was able to combat this by using infrared photography and special heat detection cameras.

'It was like some sort of crazy arms race—every time I found a way of finding them they'd try something new,' he told me.

This co-ordinated multi-agency task force has been so successful that in one of several briefings to the US House and Senate, Operation Rockfish was praised in the *Congressional Record* as 'a model for inter-agency co-operation'.

'Support from everybody, including our Governor and Attorney General, has just been great,' said Steve. 'To my knowledge, Governor Geringer [of Wyoming] is the only Governor to have flown on actual law enforcement missions. He was instrumental in getting legislation passed that makes it a felony in Wyoming to remove a fossil from State lands. In fact, at one point, he authorised moneys from the State Contingency Fund to support additional patrol flights.'

US Congresswoman Barbara Cubin has also been instrumental in pushing through new fossil legislation to better protect the heritage sites of Wyoming. Diana Ohman, former Wyoming Secretary of State, ran interference and kicked open the doors to make the first paleo-specific law enacted in the US legislature be for the State of Wyoming. US Senators Mike Enzi and Craig Thomas also played key roles in their support of Operation Rockfish and the introduction of fossil legislation in Wyoming.

Although the main thrust of the operation centred on illegal poaching of Green River fish fossils, Steve soon became aware of other sites in Wyoming that were being stripped of their fossils: fossil turtles and crocodile bones from the Bridger Formation (Eocene age) and dinosaur fossils from the Lance Creek area in eastern Wyoming, both sites on State lands. At times, Steve had to go undercover, posing as a wealthy fossil buyer to trap fossil thieves. In at least one case, he almost purchased a dinosaur skeleton taken from State land in Wyoming, but wasn't quick enough. The dealers had it out of the ground, and out of the country, within 24 hours of Steve's final bid for the specimen. He is still tracking it down, so I can't reveal any other details. He hopes to one day retrieve it for the State of Wyoming.

I asked Steve about the full extent of illegal fossil smuggling in the Green River shales of Wyoming. Roughly how many fossils were we talking about? He told me about one day on which he intercepted a large semitrailer laden with Green River shale coming down from a region that was largely State land. He stopped the truck and asked the drivers what they were carrying.

'Rocks for tiles,' they replied smugly.

Steve knew that slabs used to make tiles were laid down flat in the truck, as you can fit more in that way. This truck had rack upon rack of delicately placed vertically packed small slabs. Steve reached over and examined one of them. Sure enough, it had a fossil fish in it. Steve then did a quick estimate of the number of fossil fish slabs the truck was carrying and, knowing the current market price of the fishes, he guessed that the truck contained about US$1.5 million worth of specimens, destined for sale on the European market. The same day, on the same road, he stopped another three trucks, carrying similar loads of fish fossils. That's about US$6 million worth of fossils in one day. The haul was the result of just one season's illicit collecting from the poach sites.

'Not a bad catch, eh?'

Steve and his team had to check the origins of every fossil slab, and confiscate any which had come from State land, much to the disappointment of the dealer who had ordered the material. All the confiscated material was shared out among State collections in Wyoming museums and universities.

Steve then told me about how serious some of the individuals he was dealing with were about keeping him out of the way. He had received death threats, had seen a 'Wanted' poster with his photo on it and had a US$40 000 contract put on his head. Two of his pet cats were found one day with their necks broken.

'I guess I'm just too dumb to be scared,' was his reaction. Steve had a streak of stubbornness that would not allow him to back off. If anything, threats against him or his family only made him more determined to catch the offenders. Steve even had problems within his own agency, as some of the locals involved in the poaching activities had close relatives in the local police force.

'They applied pressure, through their family ties, to my superiors to make me lay off the cases. I told [name withheld] that if he ordered me to not enforce the law, I would not be doing my job.'

Despite all opposition, Steve continues his fight against fossil poachers.

Operation Rockfish resulted in more than 100 arrests with a 90 per cent conviction rate, in addition to the seizure of drugs, weapons and explosives, and the recovery of approximately US$7 million in fossils. Some of these recovered fossils are now on display in the Wyoming State Capital Building and at various sites throughout Washington DC, including the US Congress and the Smithsonian Institute. Some of the offenders caught in Operation Rockfish have disputed the success of the operation, claiming that it didn't result in many real arrests for fossil-related crimes. Steve told me that quite a few of the people caught and convicted in Operation Rockfish were charged with a number of felonies, and in several cases plea bargains were used to persuade defendants to plead guilty to charges deemed to be more serious than the fossil crimes. For example, if someone was charged with poaching fossils and being in possession of illegal drugs or arms, the fossil charge might be dropped. In some cases Steve had tracked down illegally-taken fossils that were being used as collateral for drug trading. The link was obvious to police and Federal agents, he told me, as once smugglers perfect their system for getting fossils out of the country, they can use that conduit to smuggle anything they like.

Steve Rogers, together with the other law enforcement agents involved in Operation Rockfish, was awarded the Military Order of the Purple Heart on 15 March 1999.

I had spent a very memorable afternoon with the Ulrichs in the town of Fossil, and as I drove through Kemmerer I kept thinking about what Steve had told me about the extent of fossil smuggling in the region. I had no idea it was on such a large scale. It made me think of the other sites around the world where abundant fish fossils can be found, and I wondered whether those countries had any means of protecting the sites and their fossils. Were there any other 'Steve Rogers' to intercept the bad guys?

We always think of large museums like the Natural History Museum in London as storehouses for enormous collections of natural history specimens taken from the four corners of the globe. In the days of the British Empire, it was common practice for scientists in the colonies to send back specimens for the great museum's collections. In the late nineteenth century, most of the Empire's colonies had no adequate means of storing and conserving fossils, so bringing them back to the museum in London was the best thing to do. These early collections have been well looked after and can now be accessed by anyone who wishes to study them. Nowadays, however, countries such as India and Australia have their own fine museums and specialist scientists who conduct research on their country's palaeontology, so new collections have been steadily building up.

The question of whether specimens should be eventually repatriated to their country of origin is a book in itself and one that I don't want to deal with here. I want to concentrate on whether laws are being enforced when fossils are sold or traded, and whether sites that are of great scientific significance can be protected from overcollecting or just plain vandalism.

The next destination in our investigation of stolen or smuggled fossils would be the Tucson show in February 2002. But to prepare myself for Tucson, I wanted to familiarise myself with the fossil legislation of other fossil-smuggling 'hot spots' around the world. South America, India and Africa were at the top of my list.

I hope that the following chapter will give some insight into the huge problems facing the authorities in those countries, who are trying to protect fossil sites of great international significance and clamp down on fossil smuggling. I can't cover every country's different problems in one small book, so I've picked out some of the worst-affected areas in terms of fossil site desecration, fossil smuggling and fossil fakery. I will also look at how one country's forward-thinking legislation is minimising fossil crime and loss of its national heritage.

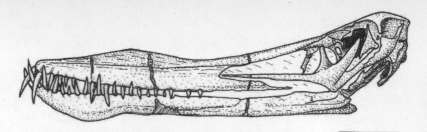

Fossil-related Crime in South America, India and Africa

11

The skull is damaged in several places due to septarian cracking of the concretion and poor preparation, mostly by the commercial fossil dealers prior to purchase. [Referring to the specimen of the skull of *Irritator*, representing a new genus of meat-eating dinosaur from Brazil] (Martill *et al.* 1996, pp. 5–6).

Sketch of the skull of the pterosaur *Anhanguera piscator* from Brazil.
(Bar scale is 10 cm)

South America

The Santana Formation in Brazil is one of South America's crowning glories of vertebrate palaeontology. Not only is there a huge diversity of exquisitely preserved fish, plant and insect fossils, but there are also scientifically important dinosaurs and pterosaurs. Specimens have been dug up and sold to local tourists and international dealers since the site's discovery in the 1820s. Only in recent years, however, have the local authorities clamped down on the illicit trade in Santana Formation fossils. Today, only a few are found on the international market, but 5–10 years ago, they were very common in rock and fossil shops all around the world.

The following is a letter I received from Dr Martha Richter, in which she answered my queries about the extent of the problems facing Brazil with regard to protection of fossil sites and fossil smuggling. Martha is a specialist in fossil fishes who is currently a visiting professor in the Departamento de Geologia/IGEO, Universidade Federal do Rio de Janeiro. Her letter is reproduced in full.

The first law dealing specifically with the protection of the fossil heritage in Brazil dated from 1942 (Decreto-Lei 4146). Since then, it is illegal to exploit fossiliferous deposits without the legal licence issued by the Departamento Nacional da Produção Mineral (DNPM), the Geological Survey of Brazil. Many other laws were also designed to prevent the export of items considered Brazilian public heritage, such as a rare fossils.

Unfortunately, fossils are still sold in many places in Brazil without much reaction from the authorities. Every now and then, the Federal Police apprehend fossils collected illegally but so far no 'de facto' condemnation of any trader has occurred. Law suit cases are very rare.

I guess that fossil trade involves many distinct aspects but fundamentally, it is a matter of philosophical stand. The principal question is 'are fossils public heritage or should they be considered private property'. In my opinion, what the ideal situation would be that all the holotypes and best specimens should be kept in museums, where they would be available for studies and the public would know about them.

The problem is that a great many new species of Santana fossils, for instance, are being described based on specimens in private collections.

What is the point of describing specimens which are not available and might be sold to somebody else who might even want to remain anonymous?

Also, the allegation that the exploitation of fossils in northeastern Brazil (Santana) helps the local economy is fallacious, because the population at large remain as miserable as ever. Only a few individuals who have access to rich buyers (who even pre-order good specimens) and the international markets, make real money.

There are problems in other areas of Brazil, not only the Araripe. I was recently involved in trying to prevent the establishment of what would be the first 'legal' enterprise to sell fossils in Brazil. Still, I'm not sure we managed to reverse the situation completely. A National Monument of the Petrified Forest was created in the State of Tocantins, but apparently, the fossil traders moved a bit to the south and I fear that they resumed activities.

Brazil is a large country, difficult to monitor. In the past, there were very few museums with qualified palaeontologists who could look after the fossils and study them. This is no longer the case, and I think there is no justification for museums around the world to buy our fossils illegally (exchange between institutions is the legal path to be explored). I would say that to make a blind eye to the illegal trade of fossils today corresponds to making a blind eye to the international trade of wild life fur or archaeological items. International co-operation and more binding treaties perhaps under the auspices of UNESCO would be very welcome initiatives.

An irritating example

Brazil's illegal fossil trade is thriving. Illegally imported fossils have been finding their way into the hands of European and American museums for many years. In 1996, Dr David Martill and his colleagues described a spectacular new skull of a crested theropod dinosaur from the Santana Formation of Brazil (Martill *et al.* 1996). The specimen had been purchased from a private dealer by the State Museum for Natural History in Karlsruhe, Germany. The new dinosaur, named *Irritator*, was represented by most of a skull which was subjected to CAT scans to reveal fantastic details of its internal features. The authors commented upon the restoration work done by the dealer who had sold the specimen:

> 'CAT scan imaging revealed that the tip of the rostrum [i.e., snout] had been artificially reconstructed to increase its length by reassembly of portions of the maxilla on to the premaxilla.' The fabrication was concealed by blocks of matrix removed from other parts of the specimen and a thick layer of Isopon™ car body filler (Martill *et al.* 1996, p. 5).

In addition, as we saw above, the skull had been damaged by the commercial fossil diggers. The importance of this specimen, for what it tells us about dinosaur biogeography, cannot be overemphasised. The authors concluded that *Irritator* is the first non-avian maniraptoran to be described from the Cretaceous of South America, and confirms the existence of a land link between South America and Asia, presumably via Africa (Martill *et al.* 1996).

Like many Brazilian fossils, *Irritator* was smuggled out of Brazil and sold on the open market. Other dinosaur remains from the Santana Formation have been seen on the market, but most are fairly scrappy. The latest dinosaur to come out of the region is *Santanaraptor placidus*, described by Alex Kellner and his colleagues (Kellner *et al.* 2002). That one is staying in Brazil.

Brazilian pterosaurs

The pterosaurs, or winged reptiles, from Santana are world famous. Some years ago Brazilian fossils were widely available, with little or no effort being made to curb their illegal export, so it was relatively easy for European museums to buy specimens, despite Brazilian laws against the loss of Brazilian heritage material. Today, there are more type specimens of Brazilian pterosaurs in European museums than in Brazil. This means that palaeontologists like Dr Alex Kellner are faced with the problem of trying to study new material without having easy access to the previously described species from his own country. Below is a list of known pterosaur species from the Santana Formation, together with the dates of the publications which described them. Most of the specimens were described soon after they were acquired, which shows us that most of them were smuggled out of Brazil from the early 1980s onwards, and sold primarily to buyers in German, Italian, Dutch and Japanese museums.

> The Departamento de Paleontologia e Geologia. Museu Naçional/ Universidade Federal do Rio de Janeiro, Brazil holds the type specimen of:
> *Araripesaurus castilhoei* (Price 1971)
> *Brasileodactylus araripensis* (Kellner 1984)
>   The following type specimens are in private collections in Brazil:
> *Anhanguera blittersdorffi* (Campos & Kellner 1985)
> *Tapejara wellnhoferi* (Kellner 1989)
> *Tupuxuara longicristatus* (Kellner & Campos 1988)
> *Ceratodactylus atrox* (Leonardi & Borgomanero 1985)
>   The Bavarian State Collections in Munich hold the type specimens of:
> *Santadactylus araripensis* (Wellnhoffer 1985)
> *Santadactylus spixi* (Wellnhoffer 1985)
> *Santadactylus pricei* (Wellnhoffer 1985)

Anhanguera santanae (Wellnhoffer 1985)
Tropeognathus mesembrinus (Wellnhoffer 1987)
Tropeognathus robustus (Wellnhoffer 1987)
Araripedactylus dehmi (Wellnhoffer 1997)

The State Museum of Natural History, Karlsruhe, Germany, holds the type specimen of:
Arthurdactylus conandoylei (Frey & Martill 1994)

The University of Amsterdam holds the type specimen of:
Santadactylus brasiliensis (de Buisonje 1980)

The Centro Studi Richerche Ligabue, in Venice, Italy holds the type specimen of:
'Cearadactylus' ligabueli (Dall Vecchia 1993)

The National Science Museum, Tokyo holds the type specimen of:
Anhanguera piscator (Kellner & Tomida 2000) and

The Iwaki Coal and Fossil Museum in Japan holds the type specimen of:
Tupuxuara leonardii (Kellner & Hasegawa 1993).

The American Museum of Natural History in New York had a large collection of Santana Formation fossils donated to it by a Dr Herbert T. Axelrod. This material mostly consisted of fishes, but there were some well-preserved pterosaurs which have been studied by Dr Alex Kellner. Dr Axelrod's generosity has led to the publication of a large volume (through his own publishing company), which provides a detailed atlas of all the known Santana Formation fossils (Maisey 1991). The Axelrod fund was also crucial for Brazilian palaeontologists such as Alex Kellner, as it allowed them to work in the American Museum of Natural History and to visit Japan to study Brazilian pterosaur and dinosaur material.

Patagonian dinosaur eggs
In late November 1998, scientists from the American Museum of Natural History, working in conjunction with

palaeontologists from Argentina, announced one of the world's most spectacular fossil discoveries from that country—beautifully-preserved nests of dinosaur eggs, some with exquisite embryos intact. The site, Auca Mahuev, is in the Patagonian desert in southern Argentina. It contains thousands of eggs, along with fossilised bits of bone, teeth and even the well-preserved skin of the dinosaur embryos, dating from around 71–89 million years old (Late Cretaceous Period). The team of scientific investigators, led by Luis Chiappe of the American Museum of Natural History and Rudolfo Coria from the Carmen Funes Museum in Argentina, published a paper in *Science* describing the exquisitely-preserved skulls of baby titanosaurids (giant, long-necked, plant-eating dinosaurs) (Chiappe, Salgado & Coria 2001). The skulls are only four centimetres in length and may shed important new light on the evolutionary history of titanosaurids through the study of their ontogenetic variations or growth changes. In his book on the Patagonian dinosaur eggs and babies (Chiappe, Dingus & Frankfurt 2001), Chiappe draws attention to the fact that although the sites are of immense scientific value to Argentina, poachers had already moved in to the area and begun digging trenches to extract the eggs. Argentina's national and provincial laws all forbid the commercial-isation of fossils, which means that all fossil material from Argentina on sale abroad has been somehow smuggled out of the country.

Argentinean palacontologists arc greatly concerned about the rising numbers of Argentinean fossils for sale on the world market. The desirability of these fossils not only encourages the criminal practices of illegal collecting, marketing, theft and smuggling, but eventually results in permanent loss of a significant part of Argentina's national fossil heritage. The first documented fossil theft from an Argentinean museum occurred in 1994. Several important specimens of Triassic reptiles were stolen from the Museum of Palaeontology at the University of La Rioja, including two skulls of *Probainognathus jenseni* (one being the holotype), the two holotype skulls of *Probelesodon lewisi* and *P. minor* and bones of the dinosaur *Riojasaurus incertus*. At the time of writing, none of the specimens has been recovered (Chure 2000).

'Collectors Corner Gardenworld', in Keysborough, Australia (see Chapter 4), claims on its Website to be 'the largest collector of dinosaur eggs in the world'. In addition to the regular range of dinosaur eggs from China, the shop recently (March 2002) advertised a titanosaurid ('saltasaur') egg from Patagonia for AU$7500. But this shop is not unique; many fossil dealers throughout the world now stock Patagonian dinosaur eggs and they are easily available over the Internet.

Perhaps one way of clamping down on illegally exported fossils being sold on the international market would be for countries like Australia to refuse import of any goods thought to be in violation of other countries' export laws. This would mean that Chinese and Patagonian dinosaur eggs, for example, would only be allowed in the country if accompanied by legitimate (and verifiable) export permits from those countries.

India

Dinosaur remains were first collected from central west India in 1922 by the famous American palaeontologist Barnum Brown. The fossils were taken back to the United States and registered into the collections of the American Museum of Natural History in New York. They included a partial skull of India's first recognised unique dinosaur, *Indosuchus raptorius*, recently redescribed in a paper by Chatterjee and Rudra (1996). The early days of fossil collecting in India were dominated by wealthy foreign collectors coming in and freely taking what they found, as was the case in most third world countries. Today India has a wealth of scientifically important sites and a number of enthusiastic scientists who collect, study and attempt to protect their country's fossil heritage.

Despite this, as in any other country where laws preventing fossil export or sales are not clear cut, the theft and smuggling of valuable fossils carries on virtually unchecked. Dinosaur eggs from Jhalod Taluka, in the Dahod district of Gujarta, about 150 km from Vadodara, are selling for 2000–20 000 rupees. Since scientists authenticated the existence of a dinosaur hatchery in Jhalod last year, palaeontologists, research students, curious foreigners and antique dealers have been visiting the site in droves. It wasn't until the *Indian Express* published a report on lax security at the site, however, that the police began to crack down and impose prohibitory orders on the removal of material from the site. Despite this, the State government's indifference to the situation and the apparent helplessness of local authorities to impose penalties on offenders means that villages in Jhalod are still being plundered. The districts of Dahod and Panchmahals share one geologist and ten public servants, but they have no official vehicle.

Local tribal groups are divided on the issue, with some acting as agents for undercover traders, others furious at the despoliation of the sites. Dinosaur eggs have always served as religious icons here and they grace almost every roadside temple but, over the past year, even these eggs have been disappearing. In some homes, children play with the eggs as exotic toys, and some locals are happy to sell the fossils to tourists, although they may say a prayer or two before completing the sale. The villagers get about US$3 an egg, three times the local daily wage, while in London sauropod eggs in silk-lined boxes sell for up to US$650 each.

A famous case of Indian fossil fraud and theft

In 1988 the world was introduced to Dr Vishwa Jit Gupta, an Indian 'palaeontologist' who had amassed an incredibly long list of scientific publications, many coauthored by eminent scientists around the world whom Gupta had contacted. The exposé by an Australian professor of palaeontology, John Talent (Talent 1995), contained damning evidence that most of Gupta's work was based on fraudulent material, whether fossils taken from other institutions but claimed to have been found in the high Himalayas of India, or photographic plates from older scientific works, which he rephotographed then reproduced to back up his claim that similar species had just been found in India. Over the course of 20 years his impressive list of scientific papers (some 400 or more works) held many anomalies in the fossil record, such as species found in an area where fossil sites simply should not occur, such as high-grade metamorphic rock zones, or species found in an area where they would be very much out of place according to prevalent biogeographic models.

Gupta began his fraudulent career in 1964, when he published two papers in the prestigious journal *Nature*, coauthored with his supervisor Professor Sahni. The papers reporting the finding of fossil graptolites in India were later found to be spurious—the graptolites they described had almost certainly been taken out of Sahni's collection of Burmese fossils. Shortly before he died, Professor Sahni discovered that his collection of Burmese graptolites had mysteriously disappeared (Talent 1995). Gupta's PhD (awarded in 1966) was also based on a mixture of fossils from unidentifiable localities and photographs plagiarised from a monograph on the fossils of the Shan States of Burma by one F.R.C. Reed. Gupta was later implicated in a number of fossil thefts and in some cases of fossil smuggling. In 1967 he spent some months working in the Geology Department of the University College of Wales, Aberystwyth, where he had free, unsupervised access to the fossil collections. It was later demonstrated that some of the Carboniferous corals, Palaeozoic conodonts and fusulinid foraminiferans which he reported from India had been taken from these collections.

In 1980, Gupta went to China and visited the Institute of Vertebrate Palaeontology and Palaeoanthropology, where he was shown a large collection of the primitive osteichthyan Devonian fish *Youngolepis praecursor* from Yunnan. In July of that year he visited Paris and, at the International Geological Congress, he showed a specimen of the fish *Youngolepis praecursor* to French palaeontologist Philippe Janvier. Insisting that the specimen came from the Himalaya region, Gupta was adamant that they should write up a paper on it immediately, as he had to return the specimen to India. Five publications (including abstracts) resulted from this collaboration, in which an 'osteolepid fish closely resembling *Youngolepis* from China' was described purportedly from Zanskar, in the Ladakh region of India. We now know that *Youngolepis* is an endemic form only found in the South China terrane (southern China and northern Vietnam), and is highly unlikely to occur in India, especially in areas where the rocks had been subjected to high pressures and are grossly deformed. This was the first case of a specimen being stolen from the Institute of Vertebrate Palaeontology and Palaeoanthropology and smuggled out of China for the sole purpose of committing scientific fraud.

After Professor Talent first exposed Gupta's misconduct, in 1992 the Panjab University Syndicate instigated an inquiry into his academic work. The final report, handed down in late April of 1994, found Gupta guilty on all charges: plagiarism, recycling fossils, inventing fictitious localities and discoveries, and misleading or duping his many coauthors. The report concluded that 'Dr Gupta is guilty of scientific mispractices' (Talent 1995). An article in the Indian weekly *The World* called for Gupta to be stripped of his PhD and DSc degrees, both of which had been demonstrated to be based upon fraudulent work. Strangely, though, when the Academic Senate of the Panjab University met to decide Gupta's fate, only five out of the 55 senators voted for his dismissal. Gupta was allowed to keep his position within the University, to supervise research students and to retain his degrees, but was barred from holding any administrative posts. Today he still works as academic at the Panjab University.

Africa

It would be fair to state that most of the fossils bought, traded or sold on the international market today come out of Africa, specifically from Morocco, a country with a population of 31 million, but a labour force of only eleven million and 23 per cent unemployment, whose main industry is the mining and processing of phosphate minerals. It was only a matter of time before fossils, the common by-product of phosphate mining, would be seen as a valuable commodity for trade and export. Today, Morocco's fossil industry is huge. Tours are conducted for avid collectors, on which they can buy specimens from the small collectors, meet with large-scale fossil dealers direct, or find fossils in the field.

Just outside the remote desert town of Erfoud is a desolate pinkish mountain called Hamar Laghad, meaning 'the Rose-Cheeked One'. This mountain, composed of layered lower Palaeozoic rocks, is full of valuable fossils. Everyone has a piece of the action here, from the small quarry operator up to the large-scale dealer and exporter. A good example is Mirzan quarry, owned by a small Arab company in Erfoud called Usine Marmar. The company sells polished fossilised table tops chiselled from the quarry for US$2000 apiece, in addition to a wide range of other fossils from the ancient (460-million-years-old) limestones. Their labourers, who spend nine hours a day sledge-hammering rock in temperatures near 50°C, are paid about US$4 a day. The desert around Erfoud is one of the richest sources of fossils in the country, especially of trilobites, an extinct group of arthropods (relatives of the insects and crabs). Erfoud could rightly be considered the trilobite capital of the world.

Palaeontologists and collectors from the West regularly visit Erfoud, in search of unusual trilobite specimens, and many new species have been bought by visiting academics. A rare trilobite can fetch sky-high prices in western fossil boutiques or on Internet auction sites. One specimen of the trilobite *Arctinurus* recently sold for US$10 000 (Chure 2000). But not all the fossils on sale in Morocco are real. Many are fakes, some so realistic that even professional palaeontologists are duped. By casting the fossil with resins mixed with actual ground rock powder, and sticking it on to a real slab of the ancient rock, the fake trilobite looks disturbingly like the real thing. There is nothing illegal about making fake fossils in Morocco. Thousands of faked trilobites are now available through the Internet, in fossil shops or on sale at large trade fairs. I have seen many in Australian fossil stores.

Despite this shady side of the industry, the Moroccan digging guild uncovers huge numbers of high-quality fossils of all ages, not just Cambrian trilobites, Ordovician nautiloids and Devonian trilobites and armoured fishes, but a huge range of crinoids, shells and corals of all ages, large fossil reptiles, in particular Cretaceous crocodiles, bits of dinosaurs, mosasaurs (marine reptiles related to today's varanids (monitor lizards)) and Eocene sharks' teeth. So abundant are the mosasaur teeth that at any fossil trade show one can select a good tooth out of box of samples, and pay only US$20. Even dinosaur teeth, such as those belonging to the giant *Spinosaurus* of *Jurassic Park 3* fame, are easily obtained. Good ones can sell for hundreds of dollars, but broken teeth can be picked up for as little as US$10. Fossilised sharks' teeth are also low-budget items, commonly found in huge numbers, and attractive to and affordable for the average small-time fossil collector.

South Africa

In late 1996 I visited a fossil collector who lived in the Karoo region of central South Africa. His name was Roy Oosthuizen, but he was known affectionately to all as Uum Roy ('Uncle Roy'). Roy had been collecting fossils on his farm and in the surrounding regions for many years and had amassed one of the country's largest private collections. He had his best specimens displayed in a separate little museum he had built at the back of his lovely home, where rows of glass cases housed a wealth of rare South African fossils. He was always helpful to visiting academics who wanted to look at his collection. Many of his prize specimens were described in scientific papers, which referred the specimens back to Roy's private numbering system.

Generous as he was with his time, Roy was very possessive of his fossils and it was only a few years before his death that he was finally persuaded to bequeath his collection to the South African Museum in Cape Town, on condition that it would remain as a separate unit within the general fossil collection. Like most private collectors, the fear of having his life's work dismantled and sold off piecemeal was enough to make him prefer that his specimens go to a State institution where not only would they be well cared for, but they would also be accessible to the public and for research purposes.

Roy's collection truly was outstanding. He had complete articulated skeletons of Triassic mammal-like reptiles such as *Lystrosaurus*, skulls of much larger predatory reptiles, the gorgonopsids, skeletons of the little Permian marine reptile *Mesosaurus* (from South Africa and South America) and a large collection of invertebrate fossils of all ages. It was in his collection that, to my delight, I saw the first Early

Devonian fish fossils found in South Africa, which have since been formally described. He also had a giant sea scorpion (eurypterid), nearly two metres long, in a giant nodule, which had been studied by Professor Waterston of the Royal Scottish Museum in Edinburgh, who had visited Roy in the early 1970s. His description of the fantastic beast was published in the *Transactions of the Royal Society of Edinburgh* (Waterston *et al.* 1985). But despite its international accolades, material in his collection could only be referred to by his own numbering system, which was inadequate for the purposes of formal taxonomic description. Still, Roy accommodated many professional palaeontologists and never forbade anyone access to his collection. His collection is today cared for by the South African Museum in Cape Town.

Fossils and fossil sites in South Africa are regarded as part of the National Estate. Fossils are deemed not to belong to individuals: they are the property of the State and, as such, are protected by law. Fossil legislation in South Africa is embodied in the *National Heritage Resources Act* (Act No. 25 of 1999) (the Act), which came into effect on 1 April 2000. The Act states that no person may destroy, damage, alter, deface, disturb, excavate, remove from its original position, collect or own, trade in or sell, export or attempt to export from South Africa, any fossil without a permit from the South African Heritage Resources Agency (SAHRA). Anyone found guilty of breaking the law is liable for a fine, several years' imprisonment, or both.

The purpose of the new legislation is not to prevent fossils from being discovered, collected and exported, but rather to ensure that the correct information is recorded and that the fossils are available in institutions for anyone to examine either now or in the future. Permits to collect fossils are normally issued only to professionally qualified palaeontologists working at museums, universities or research institutions. In some cases, individuals collecting on behalf of museums have been given permits. Every fossil collected under a permit is curated by institutions on behalf of the nation. The fossils may not be sold or given away. Even the landowner on whose land the fossils are found must have a permit to remove them from their original position and may not sell or give them to anyone other than a museum or research institution.

The Act required that any individual in possession of a fossil collection which is not the property of a museum or research/education institution, register the collection with SAHRA before 31 March 2002 (that is, two years' grace from the time of the announcement). Registration includes a binding agreement about the fate of the collection after the owner's death, to ensure that the fossils will go to a museum that can look after them on behalf of the nation. After 31 March 2002, anyone found with unregistered fossils in their possession can be prosecuted. Permit holders are required to submit to SAHRA an annual report of their collecting activities from sites that have been investigated during the year. Copies of any publications describing the fossils must also be submitted to SAHRA. Temporary Export Permits can be issued, on request, to the curators of collections to allow for collaboration with overseas workers or to arrange study loans to visiting scientists. Occasionally, fossils are exported permanently to museums or universities in other countries for display and teaching purposes, but only when there are duplicates in South African institutions.

Clearly, then, the theft and poaching of fossils has not become the problem in South Africa that it has in many other countries.

Armed with this information and experience, I was now ready to go back to the USA and participate in the world's largest fossil trade fair, at Tucson, Arizona. There I would meet up again with Steve, so that we could make enquiries about the missing Broome dinosaur footprints, and take a close look at the full extent of the legal and illegal international trade in fossils.

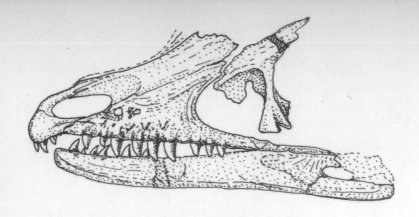

The World's Largest Fossil Fair

12

The sale of fossils is big business, bringing in about [US]$40 million annually. Those auctioned Sunday brought in [US]$160 000. The charge that science is being robbed irritates David Herskowitz, director of natural history for California-based Butterfields, a leading purveyor of such goods and the world's fourth largest auction house. 'I have a lot of things that are extremely rare, like that woolly mammoth horn,' he said in an auction preview room that included a nest of 17 raptor eggs, a 1,267.5-carat opal and two mating insects trapped in amber. 'But I do not sell anything that's crucial to science' (McFarling 2001).

Sketch of the skull of a Chinese dromaeosaurid dinosaur on sale at Tucson. The specimen is just under 5 cm in length.

The biggest deals on the international fossil trade are made in hotel and motel rooms, over beers in bars and out in the parking lots at the Tucson show. The 'Arizona Mineral and Fossil Show' brings the town of Tucson alive with dealers and buyers from all around the world. The *2002 Guide* to the show claims that 450 dealers have come to the show from all around the globe. Although most are from the USA, others have come from Germany, England, Russia, France, Italy, Belgium, Morocco, India, China, Brazil, Colombia and Australia. The main action takes place in four of the town's largest hotels, where individual dealers display their wares by day and sleep above or beside their goods by night. More entrepreneurial dealers hire expensive tables to set up larger displays in the hotel ballrooms or lobbies. Smaller operators who can't afford the expensive fees charged by the larger hotels run stalls in open-air marketplaces and parking lots or simply sell stuff from the back of their pick-ups. Not only is Tucson acknowledged as the biggest fossil and mineral trade fair in the world, it's also perhaps the most controversial. Over the past few years a number of illegally exported fossils have been sold at Tucson which later attracted worldwide media attention, the most recent of which was the highly controversial psittacosaurid skeleton from Liaoning (see Chapter 9).

Friday 1 February 2002. I arrive in Tucson at 11.20 pm, feeling more akin to my brother fossils than to the world of the living. It was a long journey from Perth, via Sydney, Los Angeles and Phoenix. I hadn't slept much at all on the flights, and had only taken one small meal, a continental breakfast on the morning flight.

Alan was there to meet me and help lug my suitcase to the waiting van. He seemed relaxed; obviously the traders had not yet reacted to him and the film crew being there to catch the action. It wouldn't take long, I thought, for word to spread and then things could become tense for us.

After all, nobody sleeps well at night if caught on camera with illegal goods, even if the legality of selling them in the USA is not in doubt. What is illegal is that some of these fossils had been smuggled out of their country of origin.

Saturday 2 February. The morning comes far too quickly for a jet-lagged palaeontologist. After reacquainting myself with bitter American coffee and a bagel, I am ready to talk turkey with Alan and Steve, who had flown in the previous day from Kemmerer in his plane, a twin-engine Cessna.

Alan wanted to go to the show and check out a few things. He'd already told me about a dealer who had a wide range of fossils from China, and asked me if I would look at them and verify their authenticity, as a lot of fossils coming out of China at the moment are fakes or composites of several different fossils.

The crew wired us with hidden microphones. Alan also carried a small digital video camera in a shoulder bag. The lens was hardly visible from the outside as the opening was barely the size of a pen top. The idea was simply to film the range of Chinese fossils on offer and try to find out from the dealers if any of the material had paperwork to show that it was a legal export. As we saw in Chapter 9, Chinese law currently prohibits the export of heritage material, such as dinosaur fossils, rare fossil birds and other specimens. As a palaeontologist who works closely with Chinese scientists, I know how hard it is to set up professional exchanges of fossil material, so getting export permits to sell fossils must be nearly impossible.

Our destination that morning was the Ramada University Inn. As we pulled in, we saw that the car park was almost full. We had stopped beforehand to turn on and check our mikes and camera, so we had about 50 minutes to play with. Alan and I walked slowly through the lobby, stopping to look at some spectacular specimens, such as a cast of the famous mummified hadrosaur with skin impressions from Canada. We sauntered through the shop area and past the pool where sausages were frying on a barbeque, the smell immediately reminding me of home for some reason. It was a nippy 10°C or so outside, so the heavy coat I was wearing to hide the microphone gear wouldn't raise an eyebrow. As I rounded the corner I noticed a small shop selling skeletons of all kinds. I had to take a closer look so we ducked in for a few minutes. To my surprise, there was a great range of human skeletons and skulls for sale, real ones. We'd have to watch our step, I thought to myself. Perhaps that's how dealers get rid of troublesome folk who pry too much into their business practices. Most of the skulls in the shop were perfectly made replicas for anatomy students. The whole range of species could be bought, including replica skulls of rare, endangered creatures such as panda or gorilla. I purchased a Moray eel skull as I was studying a fossil fish with similar adaptations. I didn't realise until later that the skull was actually a resin replica—this was clearly marked in the catalogue, but it was very realistic.

How much for the little dinosaur?

We walked into another small shop displaying a large range of Chinese fossils and immediately my eyes fell upon a small skeleton mounted in the portable glass display cabinet. It was a psittacosaurid from Liaoning Province. It wasn't from the famous Xitun Formation (Beipiao) where the feathered dinosaurs were found, but looked more like the material of *Chaoyangosaurus* from the north of the Province. The skeleton was complete and mounted in an upright posture, one hand holding a cup of coffee, the other a biscuit with a bite taken out of it.

The average price for a real, complete, mounted dinosaur skeleton can vary from about US$20 000 for a small psittacosaurid like this one, to as many as several million US dollars for something really big, popular and well-preserved (the top end of the range is represented by Sue the *Tyrannosaurus rex*, see Chapter 8).

Dinosaurs were everywhere at Tucson. In another Chinese speciality shop, a large slab of shale from the famous Liaoning bird site displayed a skeleton of a dromaeosaurid dinosaur (or 'raptor' in common dinophile vernacular). The specimen is preserved with its rigor mortis set in—tail curled up, head bent back—just like the classic first specimen of the little feathered dinosaur *Sinosauropteryx*. It had a typically dromaeosaurid type of skull, very well-preserved with well-defined, curved teeth in the jaws, some claws, leg bones and a long tail. I asked the dealer directly how he had got it out of China.

'We have many ways,' he replied with a wry grin, but offered no more details. Although it didn't have feathers attached to it like the spectacular specimen recently displayed at the American Museum of Natural History (featured in *Nature*, March 2002), it did have wisps of feather scattered around on the same slab. The skeleton appeared to be a partial, possibly a mixture of left and right counterpart pieces, but it was an impressive object and the asking price of US$68 000 was probably fair. A couple of days later one of our team went in and made a play for the specimen. She went back a few times, driving the price further and further down with each visit. On our last day in Tucson she informed us with a smile that she could have bought the feathered raptor for US$27 000 cash. Other Chinese dealers at Tucson also offered Chinese psittaco-saurid skeletons, at varying prices. One dealer who operated out of Taiwan had a range of small but relatively complete psittacosaurid dinosaur skulls starting at US$3500.

The scientific significance of this psittacosaurid material can be seen in the fact that the oldest record of this family of dinosaurs was only published in 1998 (Xu & Wang

1998). The authors describe a dinosaur from the early Early Cretaceous of the Yixian Formation of Liaoning. In December 1999, however, Dr Zhao Xijian, from the Institute of Vertebrate Palaeontology and Palaeoanthropology in Beijing, and his colleagues Cheng Zhengwu and Xu Xing published a paper on a new dinosaur from the Middle–Late Jurassic Tuchengzi Formation (140–160 million years ago) in northern Liaoning Province (Zhao *et al.* 1999). This new dinosaur, named *Chaoyangosaurus youngi*, is the world's oldest known member of the horned dinosaurs (ceratopsians), a group that would not become prominent until the late Cretaceous (65–90 million years ago). Ceratopsians included horned giants such as *Triceratops*, *Centrosaurus* and *Torosaurus*. To understand the evolution and biogeography of dinosaurs palaeontologists always search for the oldest members of a lineage and look closely at where such fossils come from. Ceratopsians have now been firmly pinpointed as originating in Asia (China, mid–late Jurassic), although one bone from Australia, attributed to *Leptoceratops* (Rich & Rich 1994), may indicate a secondary migration southwards to Gondwana.

In March 2002, another new basal ceratopsian dinosaur from Liaoning was named and described. *Liaosaurus* is the smallest true ceratopsian, as fully grown it was about the size of an average dog. It saddened me to think that some of these specimens on sale at Tucson would no doubt add new information to the story of ceratopsian origins, if they were properly studied. Most western museum curators recognise the illegality of buying Chinese dinosaurs, so it's a fairly safe bet that none of these specimens, if sold, will end up in a legitimate museum collection. Most will be sold privately, perhaps resold in Europe, but they will end up in a private collection somewhere. Some of them will never be prepared out of the rock, as they look quite attractive spread out as ancient roadkill, frozen in the gritty sands of time; the same sands and muds that, when removed, could very well provide a new piece of the puzzle in the story of ceratopsian evolution.

Yet another shop specialising in Chinese fossils was up ahead. I recognised it immediately from the huge number of Chinese dinosaur eggs stacked on racks out the front. Inside were hundreds of dinosaur eggs, many in clutches, others loose in boxes, some (possibly turtle eggs and not so perfect), for as little as US$50. Large dinosaur eggs were being sold individually for up to US$2000. However, the eggs weren't what really caught my eye. At the back of the shop was a locked glass cabinet. Inside were seven examples of the Lower Cretaceous bird, *Confuciusornis*, all from Liaoning Province. Prices ranged from a few thousand dollars to US$7500 for the best specimen, a fine example showing the wingspan and spectacular feather impressions. Alan had visited the shop the day before and expressed interest in the specimen. He told the dealer that he'd brought me along to look at it for him, to verify its quality. I carefully examined the specimen and picked up that it had three breaks through it that had been patched up with fake rock. This told me that it could possibly be a composite.

'It's certainly a good specimen,' I told Alan. 'But I'm not sure if it's been retouched or made up from other bits. There's one way to find out, though.'

I asked the dealer if he had the specimen's counterpart, the opposite side. His eyes lit up and he complimented me on my awareness of fossils, before disappearing into the rear of the shop. He returned a moment later with a small box containing a much less impressive opposite side to the fossil. It was also priced at US$7500, implying that to buy this one fossil specimen in its entirety would cost US$15 000. He had every intention of selling the counterpart as another fossil.

'I do you a special deal on both,' he said, smiling at us.

We walked away to examine the counterpart under his illuminated magnifying glass.

'But it's the one specimen,' I said to Alan. 'The part and counterpart belong together to make up one complete fossil. No museum would want a specimen like this unless both parts were together.'

My examination of the counterpart enabled me to verify that the first specimen was genuine in that none of it had been composited. The cracks had simply been filled to make it more aesthetically appealing to buyers. We discussed the quality of the specimens with the dealer and asked him where he got them from. He told us that he had an agent in Liaoning who bought the specimens from local farmers. The farmers often did the preparation on the specimens themselves, and in many cases knew how to enhance the value of specimens by adding bits and pieces from other partial skeletons. He assured us that this specimen was a genuine one, and was glad that I could confirm that it was both complete and well-prepared.

'What if we want to buy other specimens in future?' asked Alan. 'Is it possible to get more of these?'

'Yes,' he replied. 'I have a supplier in China who can get them on a regular basis.'

On examining the other six specimens, it was immediately clear that this dealer was pricing his wares purely by their marketability, or visual appeal. The scientific value of each specimen varied enormously with respect to bone preservation, degree of skeletal articulation and completeness of the feather covering on each individual. The smaller, scientifically better specimens, with the entire skull preserved and more complete feather coverings, were only US$5800, but I'm sure that an offer of US$3000–$4000 cash would have resulted in a sale. Alan questioned the dealer about the specimen's authenticity.

'If I buy this specimen, can you provide me with papers verifying what it is and where it's from?'

'Sure, no problem,' the dealer replied.

'You see,' began Alan, 'if I sell it later on to someone else back in Australia, I need to have proof that it was legally obtained. Can you provide me with some paperwork from China to verify that it's a legally exported specimen?'

At this the dealer looked suspiciously at Alan but, after a momentary pause he claimed that he could provide whatever paperwork we required. He gave each of us a colour pamphlet containing the shop's name and details and some impressive photos of specimens for sale, including the mounted little dinosaur *Psittacosaurus* from Liaoning.

'We'll go and talk about it some more,' said Alan, as we walked out.

The next shop we came to again specialised in Chinese fossils. It too had a large stock of the regular stuff, boxes of fossilised turtle and dinosaur eggs, black slabs of shale with the small nothosaur *Kueichosaurus* from the Triassic of Hunan Province, plus some very nice large specimens of the long-necked marine reptile *Sinhydrosaurus*, from Liaoning Province. These complete, articulated skeletons, some nearly 60 cm in length and all beautifully prepared, were marketed in attractive Chinese gift boxes. They were on offer from US$250–$500. This seemed like a bargain price for a complete Mesozoic reptile skeleton. We asked the shopkeeper if he had any information about the specimens. He smiled at us and took out a small photocopied image of the fossil with a few words carefully written below it in English: '*Sinohydrosaurus*, L. Jurassic, Liaoning'. On the reverse was the name of the company and its contact details. It was based in Taiwan, hence it was a legal company that was not breaking any Taiwanese laws by selling Chinese heritage fossils. Getting the fossils out of China into Taiwan was another matter, however, and this is no doubt where the risks had been taken.

An auctioneer's advice

We took a break from window-shopping to catch up with David Herskowitz, a well-known fossil auctioneer for

Butterfields in California. We wanted to ask his advice about the value of dinosaur footprints on the international market, and get his opinion on where the stolen Broome prints could possibly have ended up.

David is a real dinosaur dealer. He loves his work and is passionate and enthusiastic about his job of selling fossils. He is also very well connected in the fossil-selling trade.

What I do is I actually find the stuff, you know. I've been in this business for many years, the natural history business, so I have a lot of connections. I know a lot of people and now, because I've been doing this for eight years, most of the people know me, so usually when there's a great discovery out there, I'm one of the first people that they call.

People like to give me wish lists. I do eventually end up filling most of their needs and desires.

I mean one [client] is of a royal family you know, overseas. He's probably my top client, you know. Another top client I have is a famous publisher, you know the publishing business, real businessman executive, you know, self-made by the way. Then I have, another one of my clients is, you know, inherited all his wealth from his family and he's just a big kid actually. He's, I don't know, 62 years old, a big kid, I mean.

We asked him about dinosaur collectors.

Gosh, I have a client right now that wants a five-inch *T. rex* tooth, you know, he's been waiting four years now. He goes 'What's the largest size you've ever seen?' you know, on the market that is. I said, 'Well, usually they get as big as like maybe four, four-and-a-quarter, and that's pretty big.' 'Well, I want larger. Anything five or bigger. You come across something good five or bigger, I want it.'

And the only time you find teeth like that is usually when they're in the jaw and usually if you find the jaw you find the skull, and usually if you find the skull you . . . It's like let's go and find, buy a *T. rex*, pull out the teeth, you can throw the rest away . . . No!

163

David told me that everything Butterfields auctions must be legal, without any doubt of its authenticity. Over his eight years with the company, he has sold only two dinosaurs at auction, both relatively small ones. He said that Henry Galiano, of Mandible & Maxilla in New York, recently informed them that any fossils from South America were strictly taboo and could no longer be auctioned. Henry used to work at the American Museum of Natural History and keeps in touch with scientists there, so he was up to date with legislation issues. I'd already noticed that there were very few Brazilian fossils for sale at Tucson. Without the main buyers, museums, it seemed that the bottom had dropped out of the market.

I asked David what price he would put on the stolen stegosaur footprints from Broome. He said that fossil dinosaur footprints from the Triassic of the Connecticut Valley, *Grallator* ones, are relatively common in the USA and sell for between US$200 and US$1200, more for ones of better quality, with skin impressions, and even more if there are two or more in sequence on the one slab. The fact that the Broome specimen represents a unique fossil, the only well-preserved stegosaur prints from the southern hemisphere, makes it more valuable. Add to this the facts that they are very well-preserved, and there are two of them on the one slab, and they could fetch as much as US$10 000 to US$15 000. Of course, that's if they could be legally auctioned which, of course, they cannot. David's final advice to me was that, given time, the fossil would eventually surface. If it's in a private collection, then when the owner passes away the family will either try to sell the collection or give it to a museum. If all museums are given a detailed description of the specimen, then there's a good chance that one day in the future the footprints will be recognised and, hopefully, returned.

At the fish market

Our next stop was at one of the shops specialising in Wyoming fossils, particularly 46-million-year-old plants, fishes and birds from the Green River Formation. This particular shop had one fantastic specimen to offer— a complete fossil bird, rather large and supposedly 'undescribed' (possibly new to science), with a very reasonable asking price of a mere US$30 000. Alternatively, casts of the specimen were available for US$200. The cast on display was a very faithful reproduction of the actual specimen. I asked about its origins, and the dealer told me that it came from their own quarry, and the locality layer mentioned was one well below the regular horizons that produced the abundant fish fossils. The photos on the wall showed the excavation of the bird, verifying that it had, in fact, come from their own quarry.

I knew that fossil birds from the Green River Formation were very rare. This one looked a bit like *Gallinuloides wyomingensis* but, as I'm not a fossil bird expert and only had a brief look at it, that was a rough guess. It could well be an undescribed new species that should be studied by the fossil bird experts, so I hope it ends up in a museum collection one day. Wyoming law obligates any dealers who lease land for fossil quarrying to hand in any specimens thought to be new to science. Unfortuately, this does not apply to specimens found on private land, as was the case here.

Later that day I met with Flavio Bacchia, who has been quarrying sites in Lebanon which have produced a wide range of very nicely preserved Cretaceous fishes. Many, he told me, probably represent new genera or species as no-one has studied them yet, and many cannot easily be matched up with the previously described fauna from the Lebanon sites. These sites are famous for, among other things, complete sharks with the body outline and fins well preserved. Bacchia claimed that maybe 75 per cent of all the material he has to sell could represent new species. The specimens have all been quarried from private land, and as such there were no real legal issues with their sale.

There were a lot of fantastic fossil fishes on sale at Tucson, from all around the world, from strata of many ages. A shop specialising in Russian fossils caught my eye, as it had some superb Devonian fishes. This is my particular area of expertise, so I was compelled to go in and take a closer look. Alan was with me at the time, and he had a camera hidden in his bag. The dealer was a tall, thin Russian with a baby face, framed by wide glasses. We wandered over to examine the Devonian fish specimens. The first one that caught my attention was a perfectly preserved *Bothriolepis*, a little armoured placoderm fish with the arms (pectoral fins) preserved in three-dimensional form. It had been prepared with the arms coming out the side, like wings, the whole thing mounted on a little stand like a model aeroplane—a very nice specimen. It was priced at US$3500. No details of its locality or species were given on the label. *Bothriolepis* is a very common Devonian fish, known from over 100 different species from every continent on Earth, and as a student I had cut my teeth in fossil fish describing several of the Australian species. This, though, was one of the better ones I'd seen.

The next specimen that I was interested in was an almost complete skull (part of it had been artificially restored) of an amphibian, possibly a Triassic species. It was beautifully preserved, an opaque, whitish bone, almost like opal. Its broad triangular head showed large circular dorsal orbits, large paired nostrils and enormous holes in the palate. These features clearly defined it as a primitive amphibian, however it was labelled as 'fish', so I questioned the dealer about it. I asked him if he knew the name of the specimen and where it was from, but with a distinctly uninterested attitude, he said 'It is fish.'

'No,' I replied politely. 'It's actually an amphibian skull. Do you have the name of it somewhere?'

'No, it is a fish,' he replied sternly.

'No,' I said again. 'It is definitely not a fish. It is a fossil amphibian.'

My hackles were now well and truly raised. I know my fish fossils damn well, and this was quite definitely an amphibian skull. Alan was watching, amused at the argument that was brewing.

'No, it is definitely fish,' he huffed. At this I lost control.

'Look,' I said to him directly. 'I am a fish palaeontologist, and I have been studying fish fossils for over 20 years. This specimen is most definitely an amphibian, not a fish. Just look at the wide pterygoid vacuities in the palate here, no fish have those as large as this.'

'It is fish!' he snorted, and turned away.

I walked out. Alan was breaking up in uncontrollable laughter by this stage. He turned to me and said, smiling, 'Is fish.'

Dinosaur footprints for sale

One shop had a large sign outside it saying 'Dinosaur Footprints'. Alan and I dived for the table out front, which had an impressive array of dinosaur footprints. All of them were from the Connecticut Valley, common varieties such as *Grallator* or *Eubrontes*, and prices ranged from US$60 through to about US$250.

'Come inside, there's more here,' said the dealer as he noticed us examining the footprints.

We went in and gazed around at tables full of dinosaur eggs, nests containing several large dinosaur eggs and many other fossils. Then we looked at the other dinosaur footprints he had on display. These were mostly more of the same thing, but some were larger, better-preserved specimens. One large slab containing two big prints was on offer for around US$900.

'Do you ever get dinosaur footprints from other localities?' I asked him casually.

'No, not very often.'

'Nothing from Australia, I suppose?' I queried.

'Nope. Nothing from there.'

I then turned towards a large nest of dinosaur eggs, which I thought must have come from China. There was no label on the specimen.

'Look at this great nest of dinosaur eggs from China,' I said to Alan. On hearing my words, the dealer, who was across the other side of the room, yelled out at me, 'No those ones aren't from China. They're from Patagonia.'

We walked out and moved on.

I decided to have a look in the large display room of Canada Fossils. The first thing I saw was a magnificent skull of the hadrosaur *Edmontosaurus*, found in Alberta. The huge head had a pitiful look on its long-dead facial bones, and impressions of skin still adorned the region around the face and neck, signalling that this was no ordinary dinosaur fossil. I introduced myself to Rene Trudel, who runs the business, and he informed me that the rest of the skeleton was out the back if I wanted to take a look. It was a complete skeleton, with the mounted tail visible on display in the mirrored room behind him. I marvelled at its fantastic preservation; not only did the skin impressions continue all the way down the body to the end of the tail, but ossified ligaments held the tail stiff and horizontal. A sketch showed the rest of the skeleton, as pieces in separate large slabs, and how these all fitted together. For US$900 000, anyone could buy this magnificent dinosaur skeleton, complete with soft tissue preservation.

Yes, it was complete. *Yes* it was extraordinarily well-preserved but, unfortunately, it was just a pleasant plant-eater, not a crowd pleaser like everyone's favourite, *T. rex*. It would never fetch the same price as a perfect *T. rex* like Sue.

Rene and I discussed the fossil trade in Canada. He said that new legislation had stopped him collecting dinosaurs in Canada, so he now works with private leases in the

United States. In Canada one can get permits to sell some fossils, such as ammonites (with the beautiful polished shell known as 'korite') or petrified wood, but not vertebrate fossils.

Mike Triebold's stall had more dinosaurs for sale. Two casts of a giant new type of North American oviraptorid dinosaur, fully mounted as ready-to-display items, took pride of place in his exhibit (see Chapter 8). Mike looked resplendent in his black cowboy attire, complete with western-style hat. Mike's specimens were excavated from private land, so there were no legal squabbles over owner-ship of the specimens. The asking price for a cast of the new dinosaur was US$40 500, but they could be rented out by the month for museum displays, shopping malls, or anyone who just happened to need a dinosaur skeleton on temporary display.

On the road to Morocco

There were several shops trading in Moroccan fossils. Inside one I spied some nice examples of mosasaur (giant marine reptiles) and dinosaur teeth. There were also many beautiful examples of trilobites for sale. I told the owner that we had recently purchased some big trilobites for our museum's new 'Diamonds to Dinosaurs' Gallery, but we had bought them at bargain prices (in Australian dollars) from a local shop in Perth.

'How much would you pay for this one today?' he asked, pointing at a very large and seemingly well-preserved trilobite (a Cambrian *Paradoxides*). I was wary about it being a composite or fake, as some of the 'doctored' Moroccan specimens look so real that it's very hard to spot them, unless you saw off a corner of the specimen. The resins used to make fake fossils sometimes burn under a whirling saw blade. I told him I wasn't interested in buying the trilobite, but he persisted.

'How much would you pay today?' he asked again.

'I said no, not today, I only want a mosasaur tooth.'

'But today, how much would you want to pay for this one?' he asked yet again, looking at the magnificent trilobite.

'Well,' I replied, 'in Australia we bought a very large one for about AU$180, so if it was around US$100 it would be well priced.'

'One hundred dollars?' he said, smiling.

'Yes, that would be a fair price.'

'Good!' he said, and started wrapping the specimen in newspaper.

'But I'm not buying it today,' I repeated.

'Oh, but today's price would be how much for you?'

I walked away from him and looked at the box full of mosasaur teeth. I picked out a particularly nice one, complete with sharp edges. For US$20 it was a good specimen and it would fit well into our fossil gallery, where we only had casts of mosasaur teeth. He also had a box of broken dinosaur teeth. For another US$10 I scored a rather nice piece of spinosaurid tooth, showing the edges and the internal structure of the tooth. (Spinosaurids were fish-eating dinosaurs with long snouts. The fifteen-metre long monster *Spinosaurus aegypticus* featured as the star attraction in the *Jurassic Park 3* movie, battling with and killing a *Tyrannosaurus rex*. Ever since then, the trade in spinosaur teeth and bones has been booming.)

A large tent in the courtyard housed the Sahara Overland stall, run by Adam Abdullah Aaronson and his wife Meredith. They stocked a huge range of vertebrate fossils, pliosaur skeletons, crocodiles, mosasaur skeletons, fossil mammals, bits of dinosaur and lots of superb invertebrate fossils: the ubiquitous trilobites of all shapes, sizes and species, as well as nautiloids, a variety of crinoids, fossil corals, exotic ancient shells and so on. I browsed around but there were no fossil fishes to grab my attention.

That afternoon I had been really busy covering as much ground as possible between all the major venues. I bumped into my friend Mike Hammer. Mike is a dealer who has sold specimens to major museums throughout the USA. He asked me if I'd seen the neat Devonian fish skulls from Morocco in the motel room shop.

'No, where?' I asked him, excited by the prospect of seeing something new to science that only I might recognise.

Mike took me to the room full of Moroccan Devonian fish. A perfect little skull caught my eye, one with a high, domed neck area elaborately ornamented with elongated tubercles. It was placoderm skull from the Devonian Period, and a particularly rare variety known as a petalichthyid. As I mentally raced through all the known genera of petalichthyids (there are only a dozen or so that are well known), I was rapidly coming to the conclusion that this specimen could be a new genus. I knew at once I had to have it!

The shop had two gigantic placoderm skeletons, fully mounted. One was the giant *Titanichthys*, possibly the largest of all placoderm fishes. The other, a little smaller, was the voracious *Dunkleosteus marsaisi*. Both specimens were well-prepared (by acid etching) and were impressively mounted. I did spy a couple of loose bones mounted incorrectly and asked the man in the shop about them. When he heard that I was a fossil fish expert he was very excited and asked me about where the loose bits of bone should be fitted on the *Titanichthys* skeleton. I was still excited by my find and told him I wanted to get a good price on the petalichthyid. He rushed over to the main tent and, minutes later, in walked Adam Aaronson, who shook my hand and greeted me warmly. I showed him where the loose bones fitted on the placoderm (two were actually postmarginal bones that fitted to the side of the skull). Then I asked him about the petalichthyid skull, trying hard not to look too excited. At that point Alan came by, to drag me away to another venue. Mike told me he would talk with Adam, and see if he could get a good price for me on my find, so I told them that I would be back later, and to keep the fish skull for me!

Later that evening, I went back to Adam's shop. I had meant to meet up with Mike, to see what sort of deal he might have struck for me, but then I assumed that they would tell me what price they had negotiated. The original price was US$1500, but Adam said that for US$1000 the specimen would be mine. (He later told me that he had almost sold the specimen in Europe for over US$2400 but the sale fell through, so he decided to lower the price.) I thought hard about it, then asked him if it was quite legal, no export issues involved. Adam assured me it was a legitimate specimen, as he was one of the government-licensed fossil traders in Morocco.

'All right, I'll buy it,' I said. 'I must have it.'

I had some funds I'd raised from private sources tucked away in a museum account, so I would buy the specimen using these privately-donated funds. It would be a major acquisition for the museum's collections, as we had a world-class collection of placoderms from the Gogo sites, but no petalichthyid placoderms.

(Now that I have prepared the specimen out of its rock, I can confirm that it is different from all other known petalichthyid fish, so much so that possibly it could be in its own new family. Acid etching revealed that part of the skull had been restored with body filler and paint, but after removing these I still had the basic real skull to work on. I was very happy with my purchase.) Since then Adam has emailed me with other offers of Devonian fish skulls from Morocco, but these are of a species I know has already been described. I think my find was a lucky one; it's certainly the only time in my life I've had the thrill of walking into a shop and spotting something I know is new to science, and relatively affordable.

Undercover in Tucson

Sunday 4 February. On our second-last day in Tucson, we go undercover again, to try to find out as much as possible about smuggling operations from countries like China or Russia.

As we had spent the last two days filming openly, with cameras and boom microphones clearly in sight, we felt it would be inappropriate for Alan or me to do any of the undercover filming as our faces were now quite well-known around the venues. So instead we used Annie, a police operative who was there as part of a team investigating wider issues of fossil trading and its associated illegal activities. We wired Annie up and handed her the camera, hidden in a shoulder bag. Under Steve's guidance, her mission was to do a quick walk around the show and drop in at a number of key places we had earmarked as having significant material for sale. Her aim was simply to try to get some good footage of the specimens, and to make some simple enquiries about how the dealers managed to get their specimens into the USA. Being the second-last day of our visit we thought we had nothing to lose and could try this bolder approach. We dropped Annie off at the top of the Ramada Inn, with clear instructions to spend so many minutes at each of the designated places, so that we could follow her around in the van and listen in on her conversations. If anyone looked as if they were getting suspicious, or she felt threatened in any way, we would be only minutes away in the van.

Annie then began her walk. We also had one of our crew sauntering around the show, to keep an eye on her. All went well, no dramas, until about 40 minutes later, when Annie entered the shop of a German dealer. As soon as Annie walked in she noticed three of the guys who worked there turn and look closely at her. One of them walked around behind her and took a long look at her shoulder bag, so she immediately turned around and walked out. Outside, she whispered into her microphone that they were on to her and she needed to get away quickly. We were listening intently on the receiver.

The drill was that Annie would walk out the front of the car park, then turn left and walk two blocks down the street before we would pick her up. This gave Steve the opportunity to have another man watch her to see if anyone was following her. Later, as we were having a coffee downtown, Steve received a phone call.

'Yes, just as I suspected,' he said. 'The shop has bug detector devices in there. That's how they picked up that you had a mike.'

As soon as we had picked Annie up safely, he had sent two police operatives into the display area. They took in sophisticated bug detector sensing devices which confirmed that the shop had a bug detector device operating.

'But why would a display selling fossils go to all the trouble of having a bug sensing machine operating?' I asked Steve, somewhat naively.

'That's a very good question,' replied Steve. 'It's possible that there are bigger issues at stake, maybe smuggling or selling other goods out the back of the shop. After all, he brought about four or five guys with him from Germany. That's a lot of staff to man a small shop, isn't it?'

Monday 5 February. That afternoon the one person I'm trying to avoid walks right past me, the German dealer I had met before in Hamburg. He approaches me on the same footpath, so I cross over to avoid him, but he crosses over and walks right past me, taking a good look at me. I avoid eye contact.

After that close encounter I made a mental note to myself that I must play it cooler. It would be terrible to blow our cover now. Here was I trying to be John Long, the palaeontologist, but if he recognised me as one of the businessmen he'd met in Hamburg, then the real danger would be if he found out about Steve. If some of the dealers there who had previous run-ins with Steve knew we were working with him, it could get very uncomfortable for us, possibly dangerous.

That was the day I bought my placoderm skull from Sahara Overland, and I was feeling pretty good, so I decided to visit my friend David Ward, who was at the show buying fossil sharks' teeth for his research work. He was accompanied by Gordon Hubbell, another fossil shark's tooth expert. They were having evening drinks in Vinny's room, the same Vinny who had alerted David to the stolen *Helicoprion* fossil from Russia in 1999 (see Chapter 6). Vinny had made a good business out of collecting the fossil sharks' teeth which can be found under the sea at the famous Venice Beach site in Florida, making attractive necklaces and pendants out of them. No permits are required in Florida for collecting fossil sharks' teeth. I was warmly welcomed by the guys, a gin and tonic duly thrust into my hand as I told them about my wonderful purchase. I unwrapped the skull to show David, who took a photo of us with the specimen. Still a little apprehensive, I asked David about the legality of the Moroccan material. He assured me that Adam was okay, as he had been to Morocco to collect sharks' teeth with Adam as his guide.

I eventually went off with a group of about eight people for dinner. It was my last night in Tucson and the only thing I had to watch out for was avoiding direct contact with the Hamburg dealer, unless he recognised me first.

I was still exhilarated by my placoderm specimen and, as I walked back to the Ramada to call a taxi I couldn't stop thinking about it, hoping it would be complete with its braincase intact in the rock. The concierge told me I would have to wait 45 minutes for a cab so, in order to pass the time, I sauntered into the bar where I saw Mike Hammer. He had earlier invited me to have a beer with him, so I now saw my chance to catch up and have a chat. Japhed Boyce, whom I'd met at previous Dinofest meetings, was also there, then in came Vinny, Flash and Scott, the Florida sharks' teeth collectors. It was at that moment that I spotted the Hamburg dealer drinking a beer at the bar, about six people from where I stood. I avoided looking at him. About fifteen minutes later when I went to the bathroom he was in there washing his hands. Again I avoided eye contact and he didn't seem to recognise me.

Charlie McGovern was also there, so I chatted briefly with him, then said my goodbyes all round. I was starting to feel a little hot under the collar, so I decided to sneak out and wait for my cab outside. At that very moment, the hotel porter walked in and shouted, 'Taxi for John Long, taxi for John Long'. The words echoed around the bar. Everyone stopped chatting and looked at me. I was in a bar full of dealers, many of whom I knew to be good guys whom I'd met before, but there were others who were openly suspicious of our filming and some, like our German dealer and his friends, who might take real offence at knowing I was the man who had interviewed him in disguise in Hamburg. I felt very vulnerable at that point. None of my friends knew I was there, or could back me up if anything developed.

My martial arts training had taught me always to try to remain calm, and simply confront the problem directly. I walked straight past the guy, without looking at him, out of the bar and into my cab. Back in my hotel room, around 10.20 pm, I was getting ready for bed when the phone rang. It was one of the dealers I'd met in the bar who said that a colleague had another Devonian fish if I was interested. It was the same German dealer and it was a Russian fish fossil.

'Is it kosher?' I enquired. It was common knowledge that this man had been implicated in shifty dealings in the past, so I thought this was a very reasonable question for me to ask.

'I suppose so, but why don't you come and look at it?'

I felt uncomfortable about this. Steve had warned me that some of these dealers could have powerful connections, so his advice to me had been to steer clear of the German dealer. I decided that discretion was the better part of valour.

'Well, it sounds interesting, but I'm tired tonight, and out of here tomorrow, so tell him to send me an email. I'll look at the digital picture of it.'

After hanging up, it suddenly dawned on me that he knew where I was staying. I hadn't told anyone, even my friends from Australia, where we were staying, as Steve had told me in no uncertain terms not to let anyone know which hotel we were at. He must have found out from the hotel reception where I phoned the cab. Anyone else who was in the bar could have done the same thing. I tried to ring Alan to let him know what had happened, and that people knew where we were staying, but he was out. I slept a little uneasily that night.

In the morning, we had breakfast with Steve and said our goodbyes as he was setting off for a security posting at the Winter Olympics in Salt Lake City. I flew back to Perth. The long flight was just what I needed, time to sift through the various bits of information we had gathered on our journeys, a chance to reassess and evaluate the whole business of fossil trading and selling, site protection and legislation.

Our work at the Tucson show had kept me running on nervous energy for five days, and I had never really overcome the jet lag from the trip over. Not surprisingly, I slept well on the trans-Pacific leg, and was very pleased to get back to Perth.

The final part of the story would be to put together everything we had learned and see what other leads we could follow in the search for the missing dinosaur footprints. His help had been invaluable throughout our travels, so I was pleased and relieved when Steve rang me to say that he would be able to join us after his Olympic posting for one more trip to Broome.

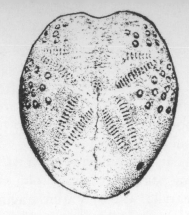

Back to Australia

13

The families of the Yawuru, Djugan and Goolarabooloo tribes are the traditional owners of the land in the shire of Broome. They are the long-standing guardians of the land and the culture of this area ... You may hear Aboriginal people referring to 'The Dreaming' or 'The Law', a very special mystery in the life of every Indigenous Australian. It is like a philosophy or theology. It is also like a Bible or sacred book, containing knowledge of the beliefs, values, ancestral beings, rituals, customs, culture, sacred sites and the languages of Aboriginal people (Broome Community and Business Directory 2000).

Sketch of a *Lovenia*, a fossilised sea urchin often found in Miocene deposits and commonly found in fossil shops. The specimen is 2.5 cm long.

We returned to Broome, to talk with Michael Latham.

Steve walked into Broome Prison and asked the guard on duty if he could talk with Michael Latham. The prison guard asked him if he knew Latham. He replied that he didn't. He was then asked the reason for his visit. Steve replied that it was a social call.

The guard kept a wary eye on Steve as he called for Michael Latham. Latham promptly arrived at the gate. Steve casually introduced himself as a US policeman who wanted to have a chat with him, adding that he might be in a position to help him. Steve then told Latham that any information he could provide about the stolen dinosaur footprints might be useful when his parole hearing came up. Furthermore, if information came to light that actually resulted in the recovery of the stolen dinosaur footprints, Steve would put in a good word to the authorities on Latham's behalf.

Initially, Latham was wary of Steve, but after a bit of casual conversation he started to let some information slip out. He said that his late brother Dennis could have been the person who carried out the first theft. He also confirmed our suspicion that the crime was committed using a chisel and plug and feather set. Steve drew a sketch of how the rock was cut away but Latham crossed it out and drew another one. His rough drawing almost exactly matched the photos we had of the crime scene.

We searched the Broome cemetery for Dennis Latham's grave and found, to our surprise, that he had passed away quite some time ago. Dennis Latham had passed away through a sudden heart attack in February 1992, leaving his wife and kids in South Australia. How could he have been implicated in the 1996 theft? The stolen footprints had been seen earlier that year by others working in the area, so we can be fairly confident that the theft did take place in or around October that year. They had first been noticed as missing by a Mr Frans Hoogland, who immediately notified the Roe family. Dennis might have been in possession of some other fossil footprints before he passed away, but they were not the stegosaur ones we were after.

Latham also mentioned that the stolen fossilised footprints could possibly have been smuggled out of Western Australia to Singapore via a man in Perth (whose name is withheld here, but his details have been passed on to the police) who regularly transported racehorses there for race meetings.

Steve asked him about Rodney Illingworth, and his involvement in the case, but Latham had nothing to say. He did mention the name of a woman who acted as a broker for Aboriginal artefacts and native art, and sometimes native fauna, out of Perth.

We weren't sure how much, if any, of Latham's information could be substantiated, but Steve seemed to think that Latham had been hard done by. Steve was of the opinion that Latham had taken the rap for the 1998 theft, on the grounds that someone higher up the chain of command had told him he and his family would be taken care of if he kept his mouth shut. When this wasn't happening after a year-and-a-half in prison, he was more than willing to share his information with a stranger, who in this case just happened to be a policeman from the USA.

Friday 5 April 2002. Alan and his crew meet with Joseph Roe at his place north of Broome. I wait back at the hotel. They return about 4.30 that afternoon, after having to track Joseph down at another site, further north than the expected meeting place.

Joseph was still not happy about any scientists, from any museum, visiting the site. The whole business of the stolen footprints had deeply grieved him. A curse had been placed on the perpetrators of the theft, and Paddy Roe was quoted in the news reports of the time as saying that the curse could also backfire on the landowners for not taking care of the site. Paddy passed away suddenly a couple of years after the theft, and Paul Foulkes, who first alerted the world to the significance of the prints' site, had also passed away in 1998 (see Chapter 2).

In his interview with Alan Carter, Joseph Roe said he had heard there was a dinosaur footprint on Dennis Latham's grave, in Broome cemetery, but it wasn't clearly visible, it was 'down in the ground'. Apparently, so the story went, Dennis had a dinosaur footprint put away, sent to South Australia and kept there for safekeeping, as he wanted it for his headstone after he died. Later that day we went to Broome cemetery, just as the sun was setting.

On finding Dennis Latham's grave, I noticed the headstone, and the one next to it (that of his father, Mick Latham Sr), were both carved out of large slabs of the ancient Broome Sandstone. Both of these slabs could, therefore, have contained a dinosaur footprint—if they did it was not visible, but it could be in the section below ground level. The actual graves were also covered in lots of small slabs of the Broome Sandstone. Close examination of these slabs suggested that the bedrock was facing downwards, exposing the underneath surface of the bedding plane. This meant that if any of these slabs were hiding a dinosaur footprint, it would be buried within the concrete that bound the mosaic of ancient rock slabs. In addition, the rock used for the graves and headstones was the red, thinly-bedded facies of the Broome Sandstone, quite unlike the rock at the stegosaurid site, so if there were a dinosaur footprint hidden here, it would most likely be one taken from the Crab Creek region, where Latham's other theft had taken place.

We left the cemetery without being any further forward in our search for the missing stegosaur prints. However, Dennis Latham's father died only a month after he did. Was this just a sad coincidence, or could it be the result of another Aboriginal curse, for stealing dinosaur footprints back in early 1991 or 1992? Michael Latham had on several occasions mentioned that his brother Dennis had been involved in the theft of dinosaur footpints. Both Dennis and Michael had worked as stone masons, with experience in cutting and removing stone slabs. Had dinosaur prints been taken from another site, and the theft not reported to the authorities? The Crab Creek region has many well-hidden dinosaur footprints, which only a few of the locals know about.

Michael Latham suggested to Alan during their last conversation that the 'first' theft of the dinosaur footprints was orchestrated by his brother, Dennis. Perhaps he could be referring to an earlier theft, not that from the stegosaurid site. When he was first questioned by Steve he had tried to implicate his brother in this theft, yet Dennis had died in early 1992. Perhaps he is too worried by retribution by the local Aboriginal people to admit to his role in the theft. According to Joseph Roe, the theft is punishable by death and the perpetrator 'can run, but he cannot hide'.

We were left with one real chance to recover the stolen footprints. The South Australian tip-off. We would take the search to Moonta.

We headed for Moonta, following Michael Latham's tip-off that there is a Broome dinosaur footprint in the backyard of a private property there. Having searched the phone book in vain for Dennis Latham's widow, we tried finding out her address through the police via John. Again, we drew a blank. There were no records that she had ever lived here.

We discussed the situation and Steve suggested that we try the 'act dumb' approach and start with the Tourist Information Centre. We went in and made polite conversation with the elderly lady at the information desk. We explained that we were having problems finding Dennis Latham's widow. Did she know her? She thought for a moment and then suggested that, 'If the lady we are looking for was pretty then maybe she had remarried after her husband's death and changed her name?' Steve smiled his big, toothy grin back at her, and said in a complimentary way, 'Well, you are still here though.' She then offered to ring someone in town who would 'know where anyone was'. They didn't, but they did suggest that we go to the Mobil Service Station and speak with a local character nicknamed 'Blackie', who apparently knew everyone in town.

Steve walked over, and asked Blackie about our quarry, but again, no go. It was as if this woman didn't have any local presence in a town where she had lived for several years. However, by a stroke of good luck, a young woman there dressed in khaki camouflage pants and T-shirt, in her early or mid-thirties, overheard Steve's conversation with Blackie and said that she knew the lady in question. Not only that, but she promptly gave Steve directions to the property. Bingo!

What should our approach be? The first option is simply to go to the property and politely ask permission to look in her yard. If she refuses, we can impress upon her that this is an important matter, and that we could ring the police for assistance if she is not willing to help us. We have a backup plan which involves John, if need be, organising for local police to come and search the premises if we can show evidence that the stolen prints might be there.

After a short drive we eventually find the cottage (we have no street number, just a description of the house and its gates to go by). We approach the front door to see if anyone's home. Steve points out to me that there are kids' dolls in the window, and a dog barks loudly at us. We peer into the back yard and see many kinds of rocks arranged around the plants, hundreds of them. No-one is home so we go back to the car and wait. Steve insists that we watch the place for an hour or so. We drive to a vantage point about a block away from the house, across a vacant paddock. From there we can watch the house without being seen. While we are waiting, Steve passes the time by asking me what can I deduce from our helpful stranger's description? I guess that maybe she is of similar age, perhaps a friend?

After two hours, a large four-wheel-drive pulls up outside the house. A man and woman get out and go into the house. We drive around to the front of the house, leap out of the car and immediately knock on the front door. Steve introduces himself and begins to chat with the woman, Michael Latham's sister-in-law. He tells her that he has spoken to Michael in Broome Prison recently, and that a rock that holds great importance for Michael could be in her garden. He explains that we are hoping to find it for him, as a favour to him. Not surprisingly, under these circumstances, she looks at us nervously but eventually invites us in and allows us to take a look around her garden, although she still seems a bit on edge. Her male companion casually looks on, but neither of them asks any more detailed questions.

I explain to them that I have a geology background and can identify rocks of all different types. I search thoroughly, but the rock isn't obviously present. The local rocks there include ancient, quartz-rich granitoids and some greenish slates from the nearby Flinders Ranges, but there was no sign of any typical Broome Sandstone. I even search up the sides and round the back of the house, peering underneath the house as well. Then, up the side laneway, under a small tree and carpeted with rotting cumquats, I see a large slab of light-coloured flat rock which could be Broome Sandstone. As I turn it over insects, ants, slaters and small black beetles scramble up my arms. I see a large central depression in the rock, and immediately think to myself, 'Bingo!' This must be it—a dinosaur footprint! After cleaning it up with my hands, however, I soon notice that the depression is merely a groove, like a wave in the rock, and in no way could be positively identified as a dinosaur footprint. I examine the matrix of the rock closely with my hand lens and, after my brain slowly reacquaints itself with my undergraduate geology notes, I conclude that it is actually an igneous rock, a coarse-grained granitoid. Such rocks do not preserve fossils, so it can't be the missing dinosaur print.

We decided it was time to get some lunch, so we cruised into Moonta. Steve was carefully watching our tail and, after a few kilometres, he drew my attention to a Commodore which had passed by us a couple of times when we were staking out the cottage. Suddenly this car was right on our tail.

'Take a quick left,' he tells me, so I turn away from the main street and pull over. We wait. Moments later the car appeared again, coming back the other way. The driver peered over and checked us out closely as the car slowly passed by.

'Yep,' Steve said smiling. 'I thought we were being tailed. Moonta's a small town. A camera crew would spark a little interest!'

'So where do we go next?' I asked him.

'Nowhere. Stay put. They can't follow us if we are going nowhere, can they?' He grinned at me.

After about ten minutes Steve told me to drive back to the Tourist Information Centre on the outskirts of town, so that we could wait a while in a safe area where lots of tourists were coming and going. We noted that the Commodore passed by us one more time and, on seeing us head out of town, its occupants probably lost interest in us. We didn't see the car again. Nonetheless, I wasn't comfortable knowing we were being followed, and was greatly relieved when it was time to leave Moonta.

I am sure the mystery of the stolen stegosaur prints and their final whereabouts will one day be resolved. They could be in a private collection which will eventually be sold off or inherited, possibly publicly displayed somewhere and recognised by a palaeontologist who has heard of the theft. Or the truth will be revealed by someone who knows where they are. There may be ways for the thief, if he is still alive, to make amends and avoid the fury of the local custodians. By returning the slab without fuss, to the exact place from where it was taken, the land may once more regain its sanctity.

The whole affair has affected many different groups. The local Aboriginal people are still very sensitive about the theft and will never allow scientists access to the site again. Dr Tony Thulborn, among others, believes the region has a lot of potential for new discoveries, and valuable work could well be ongoing if only permission were forthcoming. Local government and police now have another headache: the worry of whether any more sacred sites will be desecrated, and whether they should promote the fantastic dinosaur footprints of Broome as a tourist attraction, or keep totally quiet about them. There are difficulties in either direction.

Perhaps, one day in the future, if the stolen footprints are returned, all the different factions will be friends again. I sincerely hope so.

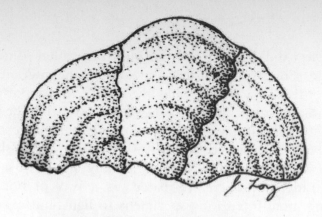

The Future of the Fossil Industry

14

Among palaeontologists and those who know, these things [Ediacaran fossils] are like ancient illuminated manuscripts.

They are relics from the very beginnings of complex life on Earth.

There's nothing like them anywhere else.

So they're incredibly sought after, extremely valuable, expensive to buy and of course illegal to take out of the State (Professor Tim Flannery, Director of South Australian Museum, ABC radio interview 19/10/00).

A sketch of my trilobite (see page 205). The specimen is 7 mm wide.

Well, we didn't end up finding the stolen dinosaur footprints, but we sure learned a lot about the international world of fossil trading. I would like to take this opportunity to summarise some of the main issues confronting the trade and commercialisation of fossils. I don't share the extreme views of some of my colleagues, who believe fossils should be the sole domain of academics, mainly because I see the fossil trade, if properly regulated, as a way of not only bringing more interesting specimens to light, but also as a potential employer of palaeontology graduates in fossil site management schemes. The stumbling block is the 'proper regulation' of the industry.

As has been shown in almost every chapter, historically, the collection and sale of fossils by amateurs to museums or private individuals has led to the discovery of important new specimens. Famous palaeontologists such as Edward Drinker Cope, a man of independent means, would collect and study fossils, publish the science and later sell his specimens to a reputable museum. The success of this system depended on the collector learning the skills of the trade, so as to extract the fossils without damage; despite the fact that the science of taphonomy (the study of how fossil deposits accumulate and of the large amount of data that can be gathered from the collecting process) was only born in the 1940s. In fact, taphonomy only took on a major role in palaeontology from the end of the 1960s.

If the fossil-trading industry is allowed to carry on unchecked, it is inevitable that many specimens will be collected regardless of crucial scientific information contained in their sites and their representation of past environments. Of course, there are many professional fossil collectors who are experts in their trade and primarily collect specimens from sites which have been thoroughly studied geologically, but only a few of these might recognise the obscure pieces of data that could be fundamentally important to science. For example, in excavating a dinosaur skeleton, the non-scientist may miss small accumulations of microfossils in and around the skeleton which tell us about any post mortem processes that occurred. Stomach contents within the sediments of the body cavity may be preserved as rock of a slightly different colour. Small teeth of scavenging animals may be hidden in the sediment near the skeleton, but unless one is actively looking for them they may not be found.

The fossil legislation now (as of 31 March 2002) adopted in South Africa has much to offer the rest of the world. It starts with the premise that all fossils belong to the State, and only registered collectors with permits may gather them from fossil sites so that they can be studied then lodged in a reputable museum or university collection. This may seem to exclude the private collector and fossil trader but, if the law was modified to accommodate professional collectors, I think it could also work well. The existing law allows for the exchange and export of South African fossils only if the specimens are well-represented in museum collections, and in that it is similar to the current *Australian Protection of Moveable Cultural Heritage Act 1992* (amended from 1986).

Those wishing to collect and sell fossils must know what they are doing, as it is the world's scientific heritage they are dealing with. In the United States, permits from the Bureau of Land Management, National Parks Service and other government bodies that manage Federal and State land resources require that fossil collectors apply for a permit, have an undergraduate degree in palaeontology and are collecting for research purposes only. What if similar restrictions applied to those who collect fossils for commercial purposes? To be registered as a collector they must have a basic qualification, not necessarily a science degree. I don't mean that dealers who have been operating successfully for years (and are already qualified at competently digging up and preparing fossils) have to be re-educated. What I have in mind is something along the lines of a brief course outlining laws and regulations pertaining to the collecting, selling and exporting of fossils from the USA and other countries. After completing such a course, there would be no excuse for a collector to plead ignorance of the laws of fossil export and cultural heritage legislation of other countries. The course could also cover issues of site protection, natural heritage and how to work with the local museum, university or geological survey. This wouldn't be a problem for those collectors who have been operating for many years, as they already know their trade well and how to work with local academic institutions. Every collector would have to have a licence to collect fossils, binding the holder to State, Federal and international laws, in order to ensure that important new sites are carefully collected and to monitor the volume of fossil material removed from each site. In turn, this would mean that dealers could be sure that all material bought and sold at international trade shows had been exported from its country of origin in full accordance with foreign cultural heritage laws.

The major problems of illegal fossil trading in the United States are caused by those dealers who disregard land management laws (poachers who steal fossils from State lands), or those who gain sole access to a significant site on private land but refuse to share the scientific process of excavation with academics interested in that site. This really scares me as a palaeontologist, that the law can place more importance on the dollar value of a site than its scientific significance (that is, what it can contribute to the whole world's total knowledge base), just because it's on private land. Does this mean, for example, that if an alien spaceship (a hypothetical object that could teach us new scientific information) is discovered buried on someone's private land, the authorities will have no say in how it is excavated or if it is destroyed because the landowner doesn't believe in spaceships? Shouldn't there be laws to protect items of the world's natural heritage found on private land or stored in private hands?

If all fossil dealers were registered, qualified and legally bound to operate within strict guidelines, this would ensure that all new scientific information is accessible to the academic community. Any fossils which are new species, or of particular scientific significance, that turn up on a commercial fossil digger's site should be donated to a major public museum. The fossil dealer should be fairly reimbursed, or given tax credits for the value of the donation. Registered expert fossil valuers would be consulted for this purpose, as we do in Australia. The Australian Tax Incentive for the Arts Scheme enables museums to acquire valuable collections through philanthropy. A private collection is professionally valued, and the valuation submitted to the taxation office. Once the valuation is accepted, tax credits are given back to the donor.

In the USA, however, some palaeontologists I know who have been working their sites for a number of years are being turned away because commercial dealers can afford to pay private landowners good fees for exploring and exploiting their sites. A system of tax credits for fossils removed should be put into place for sites that are scientifically important. The landowner enters into an agreement with the museum or university group which allows them to quarry the dinosaur bones (or trilobites or whatever). In return, the landowner will receive tax credits based on the market value of the specimens. Of course, this scheme becomes unworkable when large and highly valuable skeletons, potentially worth hundreds of thousands or even millions of dollars, are taken out of private land, as the landowner may not pay enough tax in his lifetime to reap any benefits. In such cases, the institution which is gaining such a specimen should be morally obliged to pay the landowner a reasonable percentage share from their acquisition budget. If not, landowners will simply close their doors to academic palaeontologists, and work solely with the commercial dealers who can pay them.

This raises one of the main problems with the fossil industry at present—the lack of adequate museum acquisition budgets. If museums had the funds to purchase every fantastic new specimen that comes to light (on the provision that an on-site study could be made of the skeleton, ideally before it is removed from the ground) the system could work perfectly. The value of the skeleton would then be the only issue. Most dealers add value to specimens through their preparation, as this painstaking work, that most do exceedingly well, presents the fossils in a finished, ready-to-study package. This then means that dealers can place an unrealistically high price on a specimen, way beyond the scope of most museums. There is no easy solution to this problem, except to try to educate politicians as to the real value of their country's irreplace-able natural heritage. Let's face it, most of the major art galleries or museums have far more money to buy art-works than most natural history museums have to buy fossils. A painting by Van Gogh can sell for upwards of US$25 million (for example, *Irises* was sold to the Japanese insurance company Yasuda in 1987 for £25 million), whereas the most expensive fossil ever sold was priced at US$8.36 million, including the auctioneer's fee (Sue, the *Tyrannosaurus rex*).

I wonder how much is spent each year by all the galleries in North America on acquiring art, compared with the acquisition of new scientific specimens? In my opinion, both are of equal value to society. Art nurtures our soul, gives us inspiration and makes us reflect on our humanity. Fossils fill a philosphical void about the origins and diversity of life, questions of where we have come from and where we might be going to as a species. After all, it can be argued that no single idea has had more impact on the way we think about life, in physical, metaphysical and theological terms, than Darwin's theory of evolution.

The next big issue is how to prevent the growing black market trade in smuggled fossils. Every dealer who knows his trade should by now know that all Chinese dinosaur eggs and bones are illegally smuggled out of the country, but it seems that almost everyone is turning a blind eye to this. Everyone except the law enforcement authorities in China, however, who deal out harsh penalties to those caught trying to smuggle fossils out of the country. There are two issues at stake here, one moral and one legal. Legally, fossil dealers around the world are not breaking any laws by buying and selling Chinese or Patagonian dinosaur eggs; they are operating within the laws of their own countries. Morally, they are doing the wrong thing by supporting a black-market industry which encourages local farmers back in China to raid scientific sites at random and take anything of value. This is the real crime, as the practice of stealing fossils and smuggling them out of the country could spread to many other sites. Once a successful conduit has been established for fossils to be smuggled over the border, *any* illegal or contraband material can follow the same route—this can't be allowed to continue.

The sites in central and southern China (Henan, Hubei and Guandong) which yield Chinese dinosaur eggs produce thousands of well-preserved specimens, sometimes with eggs still clustered in nests. Chinese museums do have a very good representation of these eggs in their academic collections, so the question needs to be asked, why can't some of these specimens be legally exported for sale? Why can't some of these sites be turned over to commercial collecting, if such collecting is properly supervised and regulated by the authorities? This could stamp out much of the illegal activity associated with fossil egg smuggling. I think the answer to this delicate question lies in the financial value of the black market. Chinese dinosaur eggs sell locally (here in Perth) for around AU$1200 to $7000 each. At the Tucson show, one can buy them for as little as US$300 (AU$600). Prices obviously vary enormously, as does the availability of the specimens. I can only guess that the illegality of getting them out of China makes some of them more valuable, perhaps because crooked middlemen take a share of the eggs' value in order to allow the smuggler to get them out of China. Some academics in certain institutions will, for a fee, approve export papers for 'cultural exchanges of fossils' (Simons 2001). Does this problem have an answer? Chinese cultural relic laws could be modified to allow a government-regulated trade in fossils. Thus a certain number of the dinosaur eggs, those which are well studied and well represented in public museum collections, could be sold legally on the overseas market. Alternatively, if all Chinese dinosaur eggs are deemed property of the Chinese government, then the thousands of them either in private ownership or for sale in fossil shops in nearly every country of the world would have to be returned. This of course is a ridiculous scenario.

For the present we can only acknowledge the illegality of Chinese and South American fossils on the market and prevent them from being imported into the country where they are to be sold. Customs officers should be informed what these fossils look like and if they intercept a shipment of suspicious fossils, they should contact a local museum expert to verify what they are. The specimens should be returned to the authorities of the country of origin. This will never stop the huge black market in fossil smuggling out of China, but it could act as a deterrent for potential buyers, scared of losing their goods. If the Chinese government were to issue a formal decree to the governments of each country where the trade of illegal Chinese fossils is permitted (Australia, USA, Canada, Europe, Japan), then each country would be morally obliged to seize the fossils and return them to China. As we saw in Chapter 6, this is exactly what happened in 1992 when the Australian government, acting on advice from the Federal Police, diplomatically requested the seizure of illegally exported Australian fossils in Germany. The German government acted immediately and police in Frankfurt raided a fossil dealer's premises and seized the fossils.

What is the future of the fossil dealing industry? It's clearly here to stay and, if regulated properly, will have a lot to offer the academic community. In recent times, academic palaeontology has been losing out to university administration cutbacks, and I have watched regretfully as a number of talented postgraduate students have missed out on scientific jobs and have had to turn their careers around and say goodbye to a life working with fossils. At the same time, I know of a number of significant palaeontological sites around Australia that are producing scientifically important fossils, along with quantities of other fossils of lesser significance. Such sites need on-site protection, or else they will eventually be plundered by fossil diggers, whether amateur or professionals. Ideally the sites could be developed, with a visitor centre housing an on-site ranger who regularly patrols the site, takes tour groups through for a fee and gathers up any good fossil specimens that he or she finds. State museum authorities could supervise the identification of the excess fossil material, which could then be sold to keep the visitor centre self-funded. The excess fossils would each be documented and registered, while anything new or scientifically interesting would be sent to the museum. The visitor centre could also act as an on-site field station for visiting academic field parties.

This is not to abdicate the moral responsibility that governments have to protect and conserve scientifically important sites but, with an initial injection of funds, this system could possibly be self-funding after only a few years, once ecotourism to the site picks up, and the visitor centre develops a number of educational products for sale (for example, tour packages, school educational tours, educational kits, books, pamphlets, postcards, T-shirts, fossils, fossil replicas, videos, DVDs, coffee cups . . . the list is almost endless!). If the system did become economically viable, then palaeontology graduates from around the country could be employed to manage and protect sites, develop tourism, promote education and knowledge, and undertake ongoing research.

Similar models already exist all around the world, where on-site museums take tours through their sites and sell various products to visitors. The Miguasha Park in Quebec, Canada, is a shining example of what can be achieved. In December 1999 Miguasha Park was added to the UNESCO world heritage list. The site is famous for its well-preserved Late Devonian fishes, plants and invertebrates, and for the role the specimens have played in contributing to our understanding of vertebrate evolution. In 1991 the site gained a custom-built museum which is open to the public during the warm summer months. The museum is directed by Dr Marius Arsenault, whose doctorate is in fossil fish research. Marius has been an indefatigable champion of the educational value of the fossil sites and his museum. Over 47 000 visitors to the centre each year can see the best of the Miguasha fossils on display (there are 15 000 fish fossils in their collections), visit the sites and see how the fossils are found and prepared, while learning about the fascinating geology and diverse biology of the region. School and university groups can visit for educational stays, and professional palaeontologists from around the world can use the centre for a research base. None of the Miguasha Park fossils is for sale, but replicas of some of them are available; educational products are the main commercial enterprise. The park is government-funded, and although it is not yet self-sufficient financially, it no doubt has the potential to raise a lot of tourist dollars.

And finally, what of the private fossil collector who

has no commercial interests but simply enjoys collecting fossils After discussing this issue with my colleague Dr Ken McNamara at the Western Australian Museum, we came to the conclusion that a licence, similar to a fishing or hunting licence, would be the best way to regulate private collectors in our home state. For a small annual fee the licence would allow the collector to visit fossil sites (with the landowners' permission) and take fossils for their private use. One condition of the licence would be that any material collected must be shown to an expert at the museum, who would have the authority to request the surrender of any specimens deemed to be of great scientific significance. The majority of the specimens, though, which are well-represented in State collections, would be of no interest to the museum, so a certificate of ownership could then be issued which would give the collector legal title to those fossils. The collector could then do as they pleased with them: trade, sell or export them. This system is only a suggestion. It would require considerable funds to administer, and it would create administrative positions for more expert examiners if fossil collecting keeps growing in popularity. To be foolproof, I think it would also need to be backed by a State acquisition fund whereby private collectors who find a significant specimen are reimbursed adequately for their time and the expenses involved in collecting it. Such a system currently works in Western Australia for meteorites, which all belong to the Crown, but anyone who brings in a meteorite from a remote location is entitled, under the *Museums Act*, to be recompensed for their expenses.

The various scenarios I've outlined here are just my personal views on the matter, based on my years of fossil collecting as a boy, on working in a major State-run museum as a professional palaeontologist, as someone who regularly conducts fieldwork at remote fossil sites and as someone who occasionally has had to buy fossils for our museum displays. Dealers who have been in the business for many years will obviously object to any new restrictions on their trade which could cut down their profit margins, but I hold that fossils are important primarily for the knowledge they impart to mankind, not as objects of beauty to be traded or sold.

Throughout my dealings with fossil sellers from around the world, I've come to classify them into one of two kinds. There are those who actively collect fossils, often over their entire lifetimes. Many of these collectors have a genuine love for fossils; they carefully take them out of the ground, spend many hours lovingly preparing them and want to see their best specimens end up at a major museum on public display. Peter Larson of the Black Hills Institute is a prime example of this type of dealer. Or there are also those who don't get their hands dirty but simply buy and sell fossils as commodities, the middlemen in the trade. Some of these dealers do have an interest in fossils, but others I've met seem careless of where they have come from, or what loss of scientific information may have occurred in their excavation.

Some of the stalls in Tucson had racks of fossils for display with price tags only. No information on the fossil, its age, or where it was from. Just its dollar value. I think that says it all.

While I was in Tucson, I had an idea for a way to foster more interaction between professional palaeontologists and fossil dealers. Why not have a tent or stall available for museums to offload their excess, unregistered specimens? Almost every legitimate science museum with collections makes official trades with other museums. Why not then encourage museums to bring their spare fossils or casts to the trade shows and either trade them for other fossils, or sell them to raise money for research? When I suggested this to some of the dealers at Tucson they welcomed the idea. I hope someone can put it into train for future shows.

Ultimately, professional palaeontologists need to accept that the fossil trade is here to stay, and that there are ways to work with the fossil dealers and benefit from their many years of expertise in finding and preparing fossils. Dealers need only accept the intrinsic rather than dollar value of some (not all) fossils, and respect the fact that something new to science cannot be viewed simply as a commodity to be bought or sold: it has a very special status and world heritage value. There is a middle ground here, but it will depend on institutions that buy fossils getting adequate acquisition budgets.

Epilogue: A Personal Story

In 1964 I was in grade 2, aged seven. The most eventful thing that occurred that year was that I first went hunting for fossils with my schoolmate, Desmond Matthews, and his father, a teacher who loved collecting fossils.

It was a clear spring morning, the sun shining through the remaining wisps of fog which lingered over the damp green paddocks. This was Lilydale, about 30 km from Melbourne. We set off across a paddock to a little hole in the ground about 450 metres up the hill. Magpies swooped down on us, so we had to zig-zag our way forwards to eventually reach the little hole in the ground. It was about ten metres across. There was a pond in the middle with tall reeds sticking up. The sides of the hole exposed greyish-black rock which had little orange blotches scattered throughout. Desmond's dad showed us that if we looked really closely at the little orange marks we could see the imprints of shells and other things that were the remains of sea creatures that lived millions of years ago. Trilobites were our quarry, and I was lucky enough to find several on that first trip. This was the start of my fascination with fossils.

As time passed my interest in fossils and dinosaurs grew, and I began making notes from library books, keeping lists of prehistoric animals in exercise books and sometimes making drawings of them in my spare time. My mind began to fill up with these extinct animals and soon I was able to reel off a string of strange-sounding names. My mother would often wheel me out to some tradesman fixing the plumbing, or whatever, and say 'Go on, Johnny, say those prehistoric names for the men,' to which I would respond by blurting out '*Pterodactyl, Brontosaurus, Tyrannosaurus, Triceratops, Stegosaurus*'. These little displays became common knowledge among my family and relatives, and word spread that I was a bit of a 'dinosaur expert'. I was eight at the time.

Someone suggested that I write to Professor George Baker, then the Head of the Geology Department at Melbourne University. With some help from my parents the letter was written and sent off, asking whether one of my fossils found in the hole in the ground at Lilydale was a dinosaur tooth (it looked like a picture I'd seen in a book of an *Iguanodon* tooth). Not long after, we received a nice letter from Professor Baker, inviting me to come in and meet him. I remember that day very well. My father drove me in to the University and we met the kindly old man outside the old Geology Building. In 1965, the Geology Department had a magnificent collection of fossils and minerals on display in their own museum. The first thing I learned from that meeting was that my specimen could not be a dinosaur tooth because fossils from that site were much, much older, from a time well before dinosaurs had appeared on the Earth. In fact it was a horn-shaped rugose coral, and other specimens in my collection were identified as stems of crinoids, brachiopod shells, bits of coral and parts of trilobites. The most memorable thing for me, though, was the cabinets with their thick glass tops, row after row of them, chock-full of fossils, absolutely superb examples of everything from simple life forms to real bones of dinosaurs and gigantic prehistoric mammals.

My cousin Tim Flannery and his family then lived in Sandringham and we would visit them regularly, sometimes going down to the beach at Black Rock. It was there one day, when I was about nine years old, that Tim and I found a strange heart-shaped stone on the beach. It looked like a fossil, and this was confirmed for us by the local librarian, who told us that it was a *Lovenia*, or fossil sea urchin. He told us that they were easy to pick up if we went down to Beaumaris Beach, just around the corner from Black Rock. True enough, the first time we went down there we found more *Lovenia* specimens on the beach and I also picked up a fossilised shark's tooth.

We went down there whenever we could, and occasionally met other fossil collectors who showed us their best finds. Sharks' teeth of enormous size were the most valued finds, and we were told that to find the really good stuff we would be better off snorkelling around in the shallows rather than wasting time on the beach. It didn't take long to find the underwater fossil beds and our eyes soon became adjusted to finding large slabs of fossilised whale bones, sharks' teeth, ancient fish jaws, seals' teeth, penguin bones and many other specimens of life that lived beneath the seas some seven million years ago, in the late Miocene.

It was about this time that we began taking fossils into the National Museum of Victoria to get them identified by palaeontologists Dr Tom Darragh and Ken Bell, his assistant. They would happily come down to the public enquiry area and look at our little plastic ice-cream tubs full of bones and fossil shells, and chat about what we had found. Sometimes they would take us behind the public areas into the collections, pull out a drawer and match up one of our specimens with an identical fossil. These times were very special as we had to go down into the 'dungeons', as we called them, behind the scenes, where vast cabinets were lined up in rows. Above them hung wooden dug-out canoes, or dried-up heads. This was the very soul of the museum—its collections—the results of some 150 years of collecting and curating specimens.

Our mania for collecting fossils began leading Tim and me away from Beaumaris, to any other sites that we could get to with help from our parents. We had a trip to Hamilton in western Victoria one weekend and met up with a local collector, Lionel Elmore, who showed us the best spots to collect fossils around Hamilton. Two creeks which flowed by the town, Muddy Creek and Grange Burn, had fantastic exposures of blue clay, whitish limestone and coarse gritty marl, each of which contained its own fossil assemblages. In the one area we could collect fossils of three different ages, from three different environments, and get a great range of different things. These trips were also a lot of fun—we had to walk long distances across farmland, get through barbed-wire fences, dodge charging bulls in paddocks and make dangerous creek crossings. The sort of things that kids enjoy immensely became even more fun because of the added attraction of finding fossil sharks' teeth, whale ear bones, extinct species of shells (including giant cowries) and other ancient treasures.

Other trips to Fossil Beach and Grices Creek, near Mornington in Victoria, enabled us to collect fossil shells closer to home, and occasional trips to the Geelong area, to the Waurn Ponds quarries and Fyansford Quarry at Batesford, saw our fossil sharks' teeth collection grow (I had some 500 specimens of 25 different species by the age of fourteen). My father took Tim and me on a weekend trip to Minhamite, near Hamilton, and we were allowed to stay on the farm property. During our stay we not only collected Miocene fossil sharks' teeth, shells and the like, but also

Pleistocene extinct fossil marsupials, from a younger sediment layer above the sharks' teeth bed.

In short, our fossil collections grew as we managed to sample many different sites around Victoria and became familiar with their fossil faunas. Back at school, my interest in things prehistoric continued to grow. In 1970, when I was in Form 1 of secondary school, Tim and I jointly entered the Victorian Science Talent Search Competition and won a major prize for displaying our combined fossil collections with a summary volume (about 90 pages) on the Tertiary fossils of Victoria. The prize was $50, or $25 each, a small fortune. Competition winners would also get to set up their displays at the National Museum of Victoria, for all to see and admire. Next year we entered separately and both did well. I had summarised (in two volumes totalling around 220 pages), all the fossils of Victoria, from the Cambrian trilobites through to the Pleistocene megafauna. I won the $60 major prize in the junior division (I was thirteen at the time) and had to give a speech at the prize-winning ceremony, also held in the Museum of Victoria.

The following year I entered again, with a summary of the fossil fishes of Victoria, but including description of new, unrecognised species. I won a $10 prize. I guess you can't enter fossils every year and expect to keep winning. All up, though, we both did pretty well out of the competition, and it did give me some encouragement in my desire for a palaeontological career. To that end I entered Melbourne University in 1977 and spent two years studying geology there, including completing all the third year units in palaeontology under Dr George Thomas, before transferring to complete my degree at Monash University, studying with Dr Pat Vickers-Rich. In 1980 I completed my Honours thesis on the bothriolepid placoderm fishes of Victoria, then went on to do a PhD on the Mt Howitt fish fauna. I was then faced with the hardest challenge of my life—getting a job in palaeontology.

In 1983 I was awarded a Rothmans Fellowship, not

much of a salary but an opportunity to work anywhere in Australia and do research, so I spent the next two years working in Canberra with Professor Ken Campbell and Drs Gavin Young and Richard Barwick. Those were the two most productive years of my life, and my research set me up for a Queen Elizabeth II Fellowship, so I headed over to Western Australia to have a crack at the Gogo Formation fishes. When those funds ran out, I had to look elsewhere for a job. Luckily there was a vacancy for a postdoctoral fellow at the University of Tasmania, working on a South-East Asian project under Dr Clive Burrett. The catch was that I had to work on any fossil group I was told to, but if we found fishes then I could describe them. As we did find fish fossils, in both Thailand and Vietnam, there was a lot to keep me busy. I also got to go to Antarctica for the first time that year.

The biggest break in my life came in late 1989, when I was offered a permanent position as Curator of Vertebrate Palaeontology at the Western Australian Museum. And that's where I am today.

Reflecting on those early days as a professional palaeontologist, I had a revelation. I still have the very first fossil I ever found, the trilobite from Ruddocks Quarry. It's now in the Western Australian Museum, and one day I will hand it over to them, if they really want it. Yes, I admit I'm attached to it. It's not important as a scientific specimen— it's just a poorly-preserved bug butt (pygidium). The revelation came as I was thumbing through some old copies of the *Proceedings of the Royal Society of Victoria*. I found Dr John Shergold's paper which described the trilobite fauna of the Ruddocks Quarry site at Lilydale, published in 1968, four years after I had collected my trilobites. There, in the clear photographic plates, was my trilobite pygidium. In that paper the specimens were named as a new species, *Acaste frontosa* (Shergold 1968).

So the specimens I had found in 1964 belonged to a new,

undescribed species, in fact to a genus that had not been previously recorded in Australia. As an amateur collector, I only knew the fossils as interesting-looking bits of rock that I could keep in a box at home. But they could have been of use to Dr Shergold, who at the time was studying the whole range of trilobites collected from that site.

I have told this story last of all, opening up a bit of my personal life, because the revelation that my first fossil was a new species hit me like a charging wombat. Any fossil collector who isn't a trained specialist can't fully appreciate the significance of their fossils—they often appear to be rather paltry-looking bits of what were once more complete specimens. The point is, unless collectors have some sort of professional input, both the scientific significance and the full commercial value of fossils will never be truly understood. This is not to take away from those ardent and knowledgeable collectors who do recognise the significance of most of the fossils they collect, it's only to point out an example where someone can quite easily stumble upon a significant find and not realise it. After all, I did.

We need professional palaeontologists to evaluate and signify discoveries, to place them in the big scheme of evolution, the unfolding story of life. We also need fossil hunters. Good ones may one day be the next generation of professional palaeontologists, contribute to the growing collections and new exhibitions of our major museums, or provide teaching specimens to schools and universities. But even if they don't, they may remain aficionados, lovers of fine fossils and avid collectors all their lives.

My final plea is for academics, governments and those with a commercial interest in fossils to open new lines of communication, to work together and to find a common ground where all can benefit. I hope this book is a step towards such a partnership. And to all the fossil hunters out there—keep looking. Your new species is probably just around the corner.

Appendix.
How to Check if
that Fossil is Legal*

* Legal in the moral sense. Buying a fossil that has been smuggled out of its country of origin may not be illegal in your country, but it is condoning the black-market fossil smuggling industry (which could also be condoning the smuggling of drugs or weapons), and greatly endangers fossil site protection back in that country.

Here is a quick summary of some countries' problems with regard to loss of heritage specimens in the recent years. I couldn't find information about many other countries, so if in doubt, it's best to contact the geological survey or a natural history museum in the relevant country and make enquiries.

Argentina. Argentina has a set of national and provincial laws, all of which forbid the commercialisation of fossils. All Argentinian fossil material on sale in the international market (this includes Patagonian dinosaur eggs and bones, mammalian reptiles and fossil fishes) has therefore been smuggled out of the country.

Australia.   Fossils from Australia which are sold outside the country must have an export permit from the *Protection of Moveable Cultural Heritage Act 1986* and later amendments (POMCHA) or a letter of clearance from a registered expert examiner. These letters are also sent to the POMCHA offices in Canberra so they can be checked if their authenticity is questioned. No unique new species or articulated vertebrate skeleton would be granted an export permit, so any such specimen is immediately suspect. This would also apply to any Precambrian stromatolites from the North Pole site in Western Australia; Ediacaran fossils from the Flinders Ranges; opalised vertebrate bones and some invertebrate species from Coober Pedy, Andamooka, or Lightning Ridge; any dinosaur bones; Gogo fishes; Riversleigh fossils; or any specimen from a UNESCO World Heritage listed site. *Jimbacrinus* slabs from Western Australia are often approved exports, but not if they contain new, undescribed species such as starfish fossils. Furthermore, Queensland and South Australia have their own (recent) legislation relating to collecting fossils. Before the POMCHA, fossils were prohibited exports under the 1909 *Customs Act*.

Brazil.   Since 1942 (Decreto-Lei 4146) it has been illegal to exploit fossiliferous deposits without a licence issued by the Departamento Naçional da Produção Mineral (DNPM), the Geological Survey of Brazil. A number of other laws were also enacted to prevent the export of items such as rare fossils considered Brazilian public heritage. This includes all Santana Formation pterosaurs or dinosaurs and most fishes. However, as Brazilian fossils have been collected and sold for well over 100 years, many specimens sold or exported before 1942 are quite legal. Older collections can be sold or traded legitimately, as long as it can be documented that the specimens were exported prior to 1942.

Britain. Fossils have been collected and sold in Britain for nigh on 200 years, and there are many fine old collections in circulation from which specimens are legally sold or auctioned. Most sites yielding good ammonites, ichthyosaur bones, or Old Red Sandstone fossil fishes (Scotland) are quite legal. The only problems in recent years have been with specimens taken from designated Scientific Significant Sites, such at the Lanarkshire fossil fish and invertebrate material from Scotland that was sold in Germany. Some Special Sites of Scientific Interest (SSSI) as designated by English Nature are protected, but others allow amateur collecting of surface material. If in doubt, contact English Nature to check on the status of fossils from particular sites. In general, most fossils on sale from the UK are OK.

Canada. Federal Export Permits are required under the Canadian Cultural Property Export Control List (CCPE), Group I 3—Objects recovered from the soil or waters of Canada, Palaeontological: a type fossil specimen of any value; fossil amber of any value; a vertebrate fossil specimen of a fair market value in Canada of more than $500; an invertebrate fossil specimen of a fair market value in Canada of more than $500; fossil specimens in bulk weighing 25 pounds (11.25 kg) or more of vertebrate fossils or vertebrate trace fossils of any value; fossil specimens in bulk weighing 50 pounds (22.5 kg) or more, recovered from a specific outcrop, quarry or locality, that include one or more specimens of any value of the following: invertebrate fossils, plant fossils or fossiliferous rock containing fossils of plants or invertebrates. Each State has its own fossil legislation: for example, in Alberta all fossils belong to the Crown. Individual ownership of ammonites, oyster shells, leaves and wood is permitted, but the collector must apply to the Tyrrell Museum.

China. Fossil invertebrates and plants are generally OK, except if it is something which might be in the scientifically unique catgeory (that is, a new or exceptionally rare species), or from a site that is a designated geological heritage park (for example, Chengjiang fauna). Fossil vertebrates of any kind (fish, amphibians, dinosaurs or dinosaur eggs, birds, mammals) receive the same State protection as 'cultural relics' under the cultural relics protection laws. Fossils that may have come out of China in the early days before these laws (*circa* early–mid-1980s) may be 'legally' exempt. Any Chinese vertebrate fossil for sale should have paperwork stating that it is allowed to be exported on the grounds of cultural exchange. As 'cultural exchange' relates only to the trade of fossils between major museums or academic institutions, however, this means that Chinese fossils are not to be exported for sale. Also, as 'cultural exchange' documents have been signed in the past by academics with vested interests, or forged for the buyer, it is safe to say that no vertebrate fossils coming out of China are truly legitimate. This is especially true for dinosaur eggs, and any significant vertebrate fossils from the Liaoning Early Cretaceous sites. Some of the commonest Liaoning fish fossils are permitted to be sold, and may soon be quite legal exports.

Europe.   This is not an easy one at all, because each country has its own approach to cultural heritage and export laws. Not only that but, as we discovered in Germany, each State may have different restrictions on sites and collecting. In general, only specimens from sites that have been designated as protected sites with UNESCO World Heritage status should be carefully considered. Even so, amateur collectors may have collected specimens legally from the sites prior to their World Heritage status being confirmed; these can now be resold. France has geological heritage parks, and fossils from these sites can't be collected without a permit.

Kenya.   All fossils belong to the State, with no exceptions. Anyone caught trying to take fossil material out of Kenya, especially specimens from the hominid sites around Lake Turkana, will be in serious trouble.

Mongolia. All Mongolian fossils are the property of the State. In recent years, when American and Japanese museums have led expeditions into Mongolia, it has been with the cooperation of palaeontologists from the museum in Ulan Bator. Important specimens can be studied in the USA or Japan, but only on the understanding that they will be returned to Mongolian museum collections.

Morocco. Most fossils sold out of Morocco are quite legal. The Cheftaine des Phosphates issues licences for a certain number of registered fossil collectors, who are then responsible for managing the resources and selling the specimens. Within Morocco it is a different matter, as many individuals sell fossils but are probably not licensed to do so. Buying fossils from an unlicensed seller is illegal.

Russia. The free enterprise system in Russia means that there are several thriving businesses which openly collect and sell Russian fossils. The vast majority of these specimens are legal, but the buyer should be wary of anything unique or very special that may have come out of a museum collection. A number of valuable fossils were stolen from the Palaeontological Institute in Moscow that have not yet been traced (fossil amphibian skulls, dinosaur bones from Mongolia) and are possibly still out there in the marketplace.

South Africa. As of 31 March 2002, all fossils belong to the State, with no exceptions. Permits are required for collecting, and these are only issued for academic purposes. Collectors who have private fossil collections must have had them registered with SAHRA (South African Heritage Resources Agency) by the end of March 2002, having had since 2000 to do so.

United States.   Most fossils from the USA sold on the international market are quite legal as they come from private lands. If investing in something very large and expensive, like a real dinosaur skeleton, make sure it has come from privately-owned land; not Native American Reservation land, or any form of State or government land (especially national parks and forests services lands). Each State has its own rules and regulations as to what can be collected with or without a permit, so if in doubt, it's best to check on the locality with the State authorities (the State museums or geological surveys can advise on this issue). For example, no permit is required for collecting or selling sharks' teeth from Florida. Fish fossils from Wyoming are usually fine, excepting some Green River Formation fossils that have come from State lands. Be wary of any large, impressive reptile, bird or mammal fossil for sale from the Green River Formation, because if it's from one of the State-leased quarries it has to be declared and handed in if it's something very rare, or potentially new to science. If it's from private land it can be sold without any conditions.

References

Abbott, A. 1996. Missing dinosaur skulls raise new fears of smuggling in Moscow. *Nature* 384: 499.

—1998. Moscow's 'missing fossils' come under new scrutiny. *Nature* 391: 724.

—1999. Fossil dealer charged over Russian cache. *Nature* 397: 189.

Ackerman, J. 1998. Dinosaurs take wing. The origin of birds. *National Geographic* 194: 74–99.

Boyce, J.B. 1994. Amateur and commercial collecting in palaeontology. Dinofest, Proceedings of a conference for the general public. *Paleontological Society Special Publication* 7: 99–107.

Bunk, S. 1992. Getting off on rocks. *The Bulletin*. 18 February: 56–7.

Chatterjee, S. and Rudra, D.K. 1996. KT events in India: impact, rifting, volcanism and dinosaur extinctions. Proceedings of the Gondwana Dinosaur Symposium. *Memoirs of the Queensland Museum* 39: 489–532.

Chiappe, L.M., Salgado, L. and Coria, R.A. 2001. Embryonic skulls of titanosaur sauropod dinosaurs. *Science* 293(5539): 2444–6.

Chiappe, L.M., Dingus, L. and Frankfurt, N. 2001. *Walking on Eggs: The Astonishing Discovery of Thousands of Dinosaur Eggs in the Badlands of Patagonia*. Scribner, New York.

Chure, D. 2000. New threats to old bones. The theft of fossil vertebrates from museum collections. *CRM* 5: 18–22.

Colbert, E.H. 1968. *Men and Dinosaurs*. Penguin, Middlesex, UK.

Colbert, E.H. and Merrilees, D. 1967. Cretaceous dinosaur footprints from Western Australia. *Journal of the Royal Society of Western Australia* 50: 21–5.

Dalton, R. 2001a. Elusive fossil could conceal answer to dinosaur debate. *Nature* 412: 844.

—2001b. Wandering Chinese fossil turns up at museum. *Nature* 414: 571.

Feder, T. and Abbott, A. 1994. Concern grows over 'trade' in Russian fossils. *Nature* 371: 729.

Fiffer, S. 2001. *Tyrannosaurus* Sue. Freeman & Co., New York.

Grande, L. 1984. Paleontology of the Green River Formation, with a review of the fish fauna. *Geological Society of Wyoming Bulletin* 63: 1–333.

Hawkins, T. 1834. *Memoirs of Ichthyosauri and Plesiosauri*. London.

Hecht, J. 1996. Psst, wanna buy a *Triceratops*? *New Scientist*, 14 December: 12–13.

Hou, L., Zhou, Z., Martin, L.D. and Feduccia, A. 1995. A beaked bird from the Jurassic of China. *Nature* 377: 616–18.

Kellner, A. and Tomida, Y. 2002. *Description of a New Species of Anhangueridae (Pterodactyloidea) with Comments on the*

Pterosaur Fauna from the Santana Formation (Aprian-Albian), northeastern Brazil. National Science Museum Monographs, Tokyo, 17: 1–135.

Larson, P.L. 1994. *Tyrannosaurus* sex. Dinofest, Proceedings of a conference for the general public. *Paleontological Society Special Publication* 7: 139–55.

Long, J.A. 1990. *Dinosaurs of Australia*. Reed Books, Sydney.

—1995. *The Rise of Fishes—500 Million Years of Evolution.* University of New South Wales Press, Sydney & Johns Hopkins University Press, Baltimore.

—1998. *Dinosaurs of Australia and New Zealand.* University of New South Wales Press, Sydney & Harvard University Press, Cambridge, Mass.

Long, J.A., Vickers-Rich, P., Hirsch, K., Bray, E. and Tuniz, C. 1998. The Cervantes egg: an early Malagasy tourist to Australia. *Records of the Western Australian Museum* 19: 39–46.

Maisey, J.G. (ed.) 1991. *Santana Fossils. An Illustrated Atlas.* T.F.H. Publications, Neptune City, New Jersey.

Martill, D.M., Cruickshank, A.R.I., Frey, E., Small, P.G. and Clarke, M. 1996. A new crested maniraptoran dinosaur from the Santana Formation (lower Cretaceous) of Brazil. *Journal of the Geological Society of London* 153: 5–8.

McAlpine, Alistair 1998. *Once a Jolly Bagman*. Phoenix Press, London.

McFarling, U.L. 2001. Fossils: auctioning of dinosaurs and other natural history relics angers scientists. Website: www.museum-security.org/01/017.html#5, accessed 23 January 2001.

Norell, M., Ji, Q., Gao, Yuan K., Zhao, Y. and Wang, L. 2002. Palaeontology: 'Modern' feathers on a non-avian dinosaur. *Nature* 416: 36–7.

Peters, D.S. and Qiang, J. 1998. The diapsid temprod

construction of the fossil Chinese bird *Confuciusornis*. *Senckenbergiana lethaea* 78: 153–5.

Peters, D.S. and Qiang, J. 1999. Musste *Confuciusornis* klettern? *Journal of Ornithology* 140: 41–50.

Rich, P.V. and Rich, T.H. 1994. Neoceratopsians and ornithomimosaurs: dinosaurs of Gondwana origin? *Research and Exploration* 10: 129–31.

Rowe, T., Ketcham, R.A., Denison, C., Colbert, M., Xu, X. and Currie, P.J. 2001. The *Archaeoraptor* forgery. *Nature* 410: 539–40.

Sampson, S.D., Witmer, L.M., Forster, C.A., Krause, D.W., O'Connor, P.M., Dodson, P. and Ravoavy, F. 1998. Predatory dinosaur remains from Madagascar: implications for the Cretaceous biogeography of Gondwana. *Science* 280: 1048–51.

Sampson, S.D., Carrano, M.T. and Forster, C.A. 2001. A bizarre predatory dinosaur from the Late Cretaceous of Madagascar. *Nature* 409: 504–6.

Schlosser, M. 1903. Die fossilien Saugethiere nebts einer Odontographie der recenten Antilopen. *Abhanglungen Bayerische Akademiens Wiss* 22: 1–221.

Schmidt, A.C. 2000. The '*Confuciusornis sanctus*'. An examination of Chinese Cultural Property law and policy in action. *Boston College of International and Comparative Law Review* 23 (202): 185–228.

Shergold, J.H. 1968. On the occurrence of the trilobite genera *Acaste* and *Acasteralla* in Victoria. *Proceedings of the Royal Society of Victoria* 81: 19–30.

Shu, D.G., Conway Morris, S., Han, J., Zhang, X.L, Liu, H.Q. and Liu, J.N. 2001. Primitive deuterostomes from the Chengjiang Lagerstätte (lower Cambrian, China). *Nature* 424: 419–24.

Simons, L.M. 2001. *Archaeoraptor* fossil trail. *National Geographic* October: 128–32.

Sloan, C.P. 1999. Feathers for *T. rex*? *National Geographic* 196: 98–107.

Talent, J.A. 1995. *Chaos with Conodonts and Other Fossil Biota: V.J. Gupta's career in academic fraud: Bibliographies and a short biography*. Courier Forchsunginstitut Senckenberg, Frankfurt.

Thulborn, R.A. 1990. *Dinosaur Tracks*. Chapman & Hall, London.

Thulborn, R.A., Hamley, T. and Foulkes, P. 1996. Preliminary report on sauropod dinosaur tracks in the Broome Sandstone (Lower Cretaceous) of Western Australia. *Gaia* 10: 85–96.

Tooher, J. 1998. Jamie and the elephant egg. *Australian Property Law* 6: 117–43.

Walker, C. and Ward, D. 1992. *Fossils. Eyewitness Handbooks*. Dorling Kindersley, London, New York, Stuttgart.

Waterston, C.D., Oelofsen, B.W. and Oosthuizen, R.D.F. 1985. *Cyrtoctenus wittebergensis* sp. nov. Chelicerata, Merostomata, with observations on the Scottish Silurian Stylonuroidea. *Transactions of the Royal Society of Edinburgh, Earth Sciences* 70: 251–322.

Xu, X. and Wang, X.L. 1998. New psittacosaur (Ornithischia, Ceratopsia) occurrence from the Yixian Formation of Liaoning, China and its stratigraphical significance. *Vertebrata Palasiatica* 36 (2) 81–101.

Xu, X., Xhou, Z. and Wang, X.L. 2001. The smallest known non-avian theropod dinosaur. *Nature* 408: 705–8.

Xu, X., Zhao, X. and Clark, J.M. 2001. A new therizinosaur from the Lower Jurassic lower Lufeng Formation of Yunnan, China. *Journal of Vertebrate Palaeontology* 21: 477–83.

Xu, X., Makovicky, P.J., Wang, X.L., Norell, M.A. and Hou, H.L. 2002. A ceratopsian dinosaur from China and the early evolution of Ceratopsia. *Nature* 416: 314–17.

Zhao, X.J., Cheng Z.W. and Xu, X. 1999. The earliest ceratopsian from the Tuchengzi Formation of Liaoning, China. *Journal of Vertebrate Paleontology* 19 (4): 681–91.

Acknowledgements

First of all my deepest thanks to Alan Carter and Andrew Ogilvie, the producers of the TV series, for their ideas and the financial wherewithal for getting me around to visit so many people and locations around the world. I also thank Robin Eastwood, film crew members Ian Pugsley, Laurie Chlanda and Glenn Martin for their assistance and companionship on the journey, and Sergeant Steve Rogers of the Wyoming Sheriff's department for his friendship, information and advice. Sergeant John Yates, Geraldton police, was also a great help, as was the staff of the Broome police station.

This project could not have been completed without the direct participation of the many fossil dealers, traders and private collectors who allowed themselves to be filmed and freely gave us their opinions. Wholehearted thanks to: Tom Kapitany (Melbourne), Charlie McGovern (Boulder), Mike Triebold (South Dakota), Mike Hammer, Fred and Candy Nuss (Kansas), and David Herskowitz (New York). Many international scientists also allowed themselves to be

The Dinosaur Dealers

filmed, or assisted through correspondence: Dr Bob Bakker (Boulder, USA), Dr Rupert Wild and Professor Rieschel, (Stuttgart, Germany), Dr Dieter Peters (Frankfurt, Germany), Dr Scott Sampson (Utah, USA), Dr Mark Goodwin and Dr Kevin Padian (California, USA), Dr Phil Currie (Drumheller, Canada), Dr Martha Richter (Rio de Janeiro, Brazil), Dr Billy De Klerk (Grahamstown, South Africa), Professor Meeman Zhang, Professor Qi Jiang, Dr Xu Xing and Dr Zhu Min (Beijing, China). Dr Ken McNamara is thanked for his helpful advice on fossil matters, as are David Vaughn and Robert Seleicki.

And finally, for her support through the long weeks of travel, filming and writing, I thank Heather Robinson and, for their patience, my kids, Sarah, Peter and Madeleine.

NORMANDALE COMMUNITY COLLEGE
LIBRARY
9700 FRANCE AVENUE SOUTH
BLOOMINGTON, MN 55431-4399